Key Questions in Ap
and Conservation:
A Study and Revision Guide

Key Questions in Applied Ecology and Conservation: A Study and Revision Guide

Paul A. Rees *BSc (Hons), LLM, PhD, CertEd*

CABI

CABI is a trading name of CAB International

CABI
Nosworthy Way
Wallingford
Oxfordshire OX10 8DE
UK

CABI
WeWork
One Lincoln St
24th Floor
Boston, MA 02111
USA

Tel: +44 (0)1491 832111
Fax: +44 (0)1491 833508
E-mail: info@cabi.org
Website: www.cabi.org

T: +1 (617)682-9015
E-mail: cabi-nao@cabi.org

A catalogue record for this book is available from the British Library, London, UK.

Library of Congress Cataloging-in-Publication Data

Names: Rees, Paul A., author.
Title: Key questions in applied ecology and conservation : a study and revision guide / Paul A. Rees BSc (Hons), LLM, PhD, CertEd.
Description: Boston : CAB International, 2021. | Series: Key questions | Summary: "An understanding of applied ecology and conservation is an important requirement of a wide range of programmes of study including applied biology, ecology, environmental science and wildlife conservation. This book is a study and revision guide for students following such programmes. It contains 600 multiple-choice questions (and answers) set at three levels - foundation, intermediate and advanced - and grouped into 10 major topic areas: History and foundations of applied ecology and conservation Environmental pollution and perturbations Wildlife and conservation biology Restoration biology and habitat management Agriculture, forestry and fisheries management Pest, weed and disease management Urban ecology and waste management Global environmental change Environmental and wildlife law and policy Environmental assessment, monitoring and modelling The book has been produced in a convenient format so that it can be used at any time in any place. It allows the reader to learn and revise the meaning of terms used in applied ecology and conservation, study the effects of pollution on ecosystems, the management, conservation and restoration of wildlife populations and habitats, urban ecology, global environmental change, environment law and much more. The structure of the book allows the study of one topic area at a time, progressing through simple questions to those that are more demanding. Many of the questions require students to use their knowledge to interpret information provided in the form of graphs, data or photographs"-- Provided by publisher.
Identifiers: LCCN 2020046252 (print) | LCCN 2020046253 (ebook) | ISBN 9781789248494 (paperback) | ISBN 9781789248500 (ebook) | ISBN 9781789248517 (epub)
Subjects: LCSH: Applied ecology. | Wildlife conservation.
Classification: LCC QH541.29 .R44 2021 (print) | LCC QH541.29 (ebook) | DDC 333.95--dc23
LC record available at https://lccn.loc.gov/2020046252
LC ebook record available at https://lccn.loc.gov/2020046253

References to Internet websites (URLs) were accurate at the time of writing.

ISBN-13: 978 1 78924 849 4 (paperback)
         978 1 78924 850 0 (ePDF)
         978 1 78924 851 7 (ePub)

Commissioning Editor: Ward Cooper
Editorial Assistant: Emma McCann
Production Editor: James Bishop

Typeset by SPi, Pondicherry, India
Printed and bound in the UK by CPI Group (UK) Ltd, Croydon, CR0 4YY

Contents

About the Author

Paul Rees was a senior lecturer in the School of Science, Engineering and Environment at the University of Salford, United Kingdom, until his retirement in 2020 after 22 years. He holds a PhD in animal ecology and behaviour from the University of Bradford. Paul previously lectured at three Further Education Colleges and a Higher Education College in the UK, and trained biology teachers at Sokoto College of Education in Nigeria. He has taught from GCE 'O'/GCSE level to MSc level and has been an external examiner for a range of taught programmes, from Higher National Diploma to MSc level, at six British universities. Paul has published papers on mammal behaviour and ecology, wildlife law, and the role of zoos in conservation, along with eight textbooks concerned with ecology, zoo biology, wildlife law and elephants. He is the author of *Key Questions in Ecology: A Study and Revision Guide*.

Preface

The world is facing a very wide range of ecological challenges. Many have been around for some time, such as air, water and land pollution, urbanisation and soil erosion. Others are now emerging, such as the risk of global zoonotic pandemics, rapid biodiversity loss and global climate change.

The modern environmental movement began in the 1970s and helped to bring applied ecology and conservation to the attention of the general public. New disciplines have emerged including conservation biology, restoration ecology and urban ecology and these have spawned new courses in universities all over the world.

In recent years many college and university courses have been adopting multiple-choice questions (MCQs) as a method of assessment. This trend and the sudden switch to online teaching and learning at many institutions as a result of the social distancing necessary to slow the spread of the COVID-19 pandemic have created a unique opportunity for books of MCQs to add value to many science courses.

No two courses on applied ecology or conservation are the same. I have attempted to produce a book that covers many of the topics that I would expect to be included in courses of this nature along with basic facts, including historical facts, that I think it would be useful for students to know. While this book is primarily intended for students of applied ecology and conservation I hope some people who have a general interest in ecology will find this book a useful introduction to some of the history, development, terminology and key concepts of the subject.

Acknowledgements

I am grateful to Ward Cooper (Commissioning Editor) and his colleagues at CABI for their encouragement and support during the production of this book.

Several of the figures used in this book have been published elsewhere and I am grateful to my various publishers for permission to reuse them here. They are Figs. 1.2 and 4.2 (Wiley-Blackwell) and Figs. 2.1, 3.1, 5.2 and 6.3 (Elsevier). Figs 1.3 and 1.4 were made available by the United States Library of Congress.

My daughter kindly found time to check the manuscript while balancing work and family commitments and I am very grateful for this. Any errors that remain are, of course, mine.

This book was produced during the COVID-19 pandemic by people in the United Kingdom and India who were working under very challenging conditions and I thank them all for their efforts.

Finally I would like to thank my wife and daughter for understanding that from time to time over the last few months I have needed to prioritise writing over fully engaging with assorted family activities. They would, however, have to acknowledge that I did make time to watch episodes of *Thunderbirds* with my grandson.

How to Use This Book

The questions are arranged by topic and divided into three levels: foundation, intermediate and advanced. These levels are not intended to reflect any particular curriculum but rather general levels of difficulty, and should not be taken too seriously. Knowledge of definitions and basic facts are dealt with at the foundation level; the intermediate level contains questions on methods and processes; and the advanced level contains questions involving more obscure facts and definitions, along with calculations, and mathematical concepts where appropriate. However, there is some variation between chapters as not all of the topics covered lend themselves to this approach. Students are advised to check the syllabuses they are following in detail before relying too much on this book as a preparation for specific exams.

Students are encouraged to complete a whole chapter – or at least a complete section (foundation, intermediate or advanced) – before looking at the answers. This is because the explanations for some answers may assist in selecting the correct answer to subsequent questions, although I have tried to avoid this where possible. The order in which the chapters are attempted does not really matter because each is about a different area of applied ecology or conservation. However, within any chapter you are advised to attempt the foundation questions first, followed by the intermediate questions and finally the advanced questions. Some of the questions involve calculations and it would be useful to have access to a calculator when attempting them.

History and Foundations of Applied Ecology and Conservation

1

This chapter contains questions about the history of applied ecology and conservation along with others concerned with some basic principles.

Foundation

1.1f **Ernst Haeckel was a German zoologist who first used the term**

    a. ecosystem

    b. ecology

    c. applied ecology

    d. pollution

1.2f **The term 'biodiversity' was first used in the**

    a. 1960s

    b. 1970s

    c. 1980s

    d. 1990s

1.3f **The oldest national park in the world is**

    a. the Serengeti National Park, Tanzania

    b. the Peak District National Park, England

    c. Banff National Park, Canada

    d. Yellowstone National Park, United States

1.4f A sacred grove is an area of religious and cultural importance in some societies that affords protection to wildlife. It consists predominantly of

 a. grassland

 b. woodland

 c. marshland

 d. fresh water

1.5f The original focus of the World Wildlife Fund (now the World Wide Fund for Nature) was on the plight of

 a. East African game animals

 b. Indian wildlife

 c. whales and dolphins

 d. the Amazon rainforest

1.6f Biophilia is

 a. the scientific name for a fear of animals

 b. a type of naturalistic zoo enclosure design

 c. the scientific name of a group of rare amphibians

 d. the emotional affiliation of human beings to other living organisms

1.7f The extinct bird shown in Fig. 1.1 is

Fig. 1.1.

a. a bush moa (*Anomalopteryx didiformis*)

b. a dodo (*Raphus cucullatus*)

c. an elephant bird (*Aepyornis maximus*)

d. a crested moa (*Pachyornis australis*)

1.8f **Which of the following could not be considered one of the founders of the modern environmental movement?**

a. Barry Commoner

b. Rachel Carson

c. Paul Ehrlich

d. Ralph Nader

1.9f **Which of the following categories of protected area defined by the International Union for the Conservation of Nature (IUCN) would you expect to experience the lowest levels of human presence and activity?**

a. National park

b. Wilderness area

c. Protected landscape

d. Habitat management area

1.10f **The Swedish schoolgirl Greta Thunberg is famous for encouraging young people to lobby governments to take action on**

a. biodiversity loss

b. plastic pollution of the oceans

c. climate change

d. excessive product packaging

1.11f **The Gaia hypothesis suggests that the Earth may be studied as a single functioning organism and was proposed by**

a. Paul Ehrlich

b. Norman Myers

c. Carl Sagan

d. Aldo Leopold

1.12f Complete the following sentence by selecting the most appropriate word from the list below: 'The .... use of a resource ensures that it is not over-exploited.'

   a. sustainable

   b. predictable

   c. quantifiable

   d. managed

1.13f The book written by the American biologist Rachel Carson that alerted the world to the damage done to nature by chemical insecticides and other agricultural chemicals was called

   a. *Silent Winter*

   b. *Silent Spring*

   c. *Silent Summer*

   d. *Silent Autumn*

1.14f The Green Revolution that began in North America and Western Europe in the 1940s concerned

   a. the development of drought-resistant trees

   b. improvements in the protection of the world's forests

   c. the development of new high-yielding breeds of rice, wheat and maize

   d. the development of heavy metal-tolerant grasses for use in reclaiming derelict land

1.15f Match the branches of ecology with the correct definitions in Table 1.1.

   a. A

   b. B

   c. C

   d. D

Table 1.1

Definition	A	B	C	D
Examines the relationship between a society and its natural environment	Political ecology	Social ecology	Political ecology	Cultural ecology
Examines how biodiversity research informs public policy	Cultural ecology	Cultural ecology	Social ecology	Political ecology
Examines the relationship between ecological problems and social issues	Social ecology	Political ecology	Cultural ecology	Social ecology

1.16f John Muir was an important figure in the protection of

   a. landscapes

   b. cetaceans

   c. tropical forests

   d. African wildlife

1.17f Who was primarily responsible for the creation of the World Wildlife Fund (now the World Wide Fund for Nature)?

   a. Sir Peter Scott

   b. Sir Roger Tory Peterson

   c. Sir David Attenborough

   d. Dr Desmond Morris

1.18f Due to the dominant human influence on the Earth the current geological age has come to be known by some scientists as the

   a. Anthroassic

   b. Anthropocene

   c. Anthovian

   d. Anthrozoic

1.19f Which organisation operates the ship in Fig. 1.2?

    a. World Wide Fund for Nature

    b. Extinction Rebellion

    c. Friends of the Earth

    d. Greenpeace

Fig. 1.2.

1.20f **Dr Wangarï Maathai won the Nobel Peace Prize in 2004 for her efforts to protect the environment, especially**

a. planting trees in Kenya

b. protecting wildlife in Uganda

c. establishing an environmental education programme in South Africa

d. campaigning against water pollution in Mozambique

Intermediate

1.1i **The study of indigenous peoples and how they use wild plants is known as**

a. anthrobotany

b. ecobotany

c. ethnobotany

d. homobotany

1.2i **Which was the first national park to be designated in the United Kingdom?**

a. Snowdonia

b. Lake District

c. North York Moors

d. Peak District

1.3i **Who wrote *Walden*, a book published in 1854, that was one man's reflection on simple living and self-reliance in natural surroundings?**

a. Henry David Thoreau

b. Ralph Waldo Emerson

c. Walt Whitman

d. John Muir

1.4i The Imperial Bureau of Entomology was established in 1913 to develop methods of controlling insect pests in the colonies of

a. France

b. Germany

c. Great Britain

d. Portugal

1.5i Sites in North America where large numbers of bison (*Bison bison*) were driven over cliffs to their death by native Americans as a method of hunting are called

a. bison drops

b. buffalo traps

c. bison falls

d. buffalo jumps

1.6i In 1936 the Great Plains Committee presented President Franklin D. Roosevelt with a report on *The Future of the Great Plains* that argued that the Dust Bowl of the American prairies (Fig. 1.3) was

Fig. 1.3.

a. caused by climate change

b. caused by human activity

c. a natural feature of the ecosystem

d. caused by the cessation of grass burning by native Americans

1.7i **In which year did a 'pea soup' fog of industrial pollutants envelop London for three weeks resulting in the deaths of 4000 people from respiratory illness and changes to air pollution law?**

a. 1949

b. 1952

c. 1961

d. 1965

1.8i. **Fauna and Flora International (FFI) publishes a journal of international conservation entitled**

a. *Oribi*

b. *Impala*

c. *Oryx*

d. *Eland*

1.9i **The IUCN Red List is an inventory of**

a. endangered species of organisms

b. forests whose future is threatened by logging and agriculture

c. toxic chemicals that contaminate the marine environment

d. indigenous tribes whose existence is threatened by habitat destruction

1.10i **Which environmental organisation, founded in 1971, is famous for campaigns against whaling and nuclear power?**

a. Greenpeace

b. Extinction Rebellion

c. Earthwatch International

d. Friends of the Earth

1.11i When and where was the environmental organisation the Sierra Club founded?

    a. London, 1910

    b. San Francisco, 1892

    c. New York, 1923

    d. Amsterdam, 1899

1.12i *Martha* was the name given to an animal that died in 1914 at the Cincinnati Zoo in the United States. She was the last specimen of the

    a. quagga (*Equus quagga quagga*)

    b. Tasmanian tiger (*Thylacinus cynocephalus*)

    c. passenger pigeon (*Ectopistes migratorius*)

    d. Carolina parakeet (*Conuropsis carolinensis*)

1.13i The first 'Earth Day' was celebrated in

    a. 1960

    b. 1970

    c. 1980

    d. 1990

1.14i Who published a book in 1968 entitled *The Population Bomb* about the growth of the human population and the possibility of world famine?

    a. Carl Sagan

    b. Ralph Nader

    c. Eugene Odum

    d. Paul Ehrlich

1.15i Studies of the development of human attitudes towards the environment, the reasons why people value nature and the behaviour of people towards nature are likely to be undertaken by someone with expertise in

   a. conservation psychology

   b. environmental ethics

   c. environmental economics

   d. behavioural ecology

1.16i Which of the following countries has the highest number of endemic species of mammals, birds and amphibians?

   a. Ecuador

   b. Australia

   c. Madagascar

   d. India

1.17i Which of the following zoogeographical realms would you expect to have the greatest number of butterfly species?

   a. Nearctic

   b. Ethiopian

   c. Palaearctic

   d. Neotropical

1.18i When ranchers graze their animals on a common field each will want to maximize his profits by increasing the size of his herd. If each rancher follows this path without regulatory controls the field will eventually be damaged by overgrazing. Although each individual would be acting in his own self interests, collectively they would be behaving contrary to the common good by damaging a common resource. This exemplifies the problem known as the

   a. catastrophe of common resources

   b. disaster of common assets

   c. tragedy of the commons

   d. calamity of common land

1.19i   Which of the following is least likely to engage in *ex-situ* conservation projects?

   a. A zoological gardens

   b. A botanical gardens

   c. A public aquarium

   d. A national park

1.20i   Who formulated the 'Four Laws of Ecology'?

   a. Eugene Odum

   b. Barry Commoner

   c. Robert MacArthur

   d. Edward Wilson

Advanced

1.1a   The ethical position of some conservationists whereby non-human life, mountains and rivers are respected and given moral rights and low consumption, social decentralisation and connectedness are promoted, is called

   a. broad ecology

   b. profound ecology

   c. deep ecology

   d. wide ecology

1.2a   In 1973 a grass roots protest movement was formed in northern India called 'Chipko Andolan' or the 'Hugging Movement' whose aim was to protect

   a. tigers

   b. elephants

   c. soil

   d. trees

1.3a **Which of the following tables (Table 1.2) correctly pairs environmental campaigners with their causes?**

a. A

b. B

c. C

d. D

Table 1.2

A		B	
Chico Mendes	Amazon rainforest destruction	Karen Silkwood	Nuclear pollution
Dian Fossey	Nuclear pollution	Chico Mendes	Amazon rainforest destruction
Richard Leakey	Mountain gorillas	Dian Fossey	Mountain gorillas
Karen Silkwood	Kenyan wildlife	Richard Leakey	Kenyan wildlife
C		D	
Chico Mendes	Nuclear pollution	Richard Leakey	Nuclear pollution
Dian Fossey	Kenyan wildlife	Chico Mendes	Amazon rainforest destruction
Richard Leakey	Mountain gorillas	Dian Fossey	Mountain gorillas
Karen Silkwood	Amazon rainforest destruction	Karen Silkwood	Kenyan wildlife

1.4a **The first international conference concerned with the global effects of environmental pollution and destruction was held in Stockholm in 1972. It was the**

a. United Nations Conference on the Human Environment

b. United Nations Conference on the Environment

c. United Nations Conference on Human Ecology

d. United Nations Conference on the Future of the Environment

1.5a **A report published in 1972 on the results of a computer simulation of exponential economic and human population growth with a finite supply of resources was entitled**

a. *Future Sustainability*

b. *The Limits to Growth*

c. *Finite Resources*

d. *Reframing Economics*

1.6a **The term 'biosphere' was coined by the**

a. mineralogist Vladimir Ivanovich Vernadsky

b. zoologist Ernst Haeckel

c. mineralogist Abraham Gottlob Werner

d. botanist Joseph Dalton Hooker

1.7a **Which former President of the United States (Fig. 1.4) wrote a book entitled *African Game Trails* and sent a large number of animal specimens to the National Museum of Natural History following a safari that began in Mombasa, Kenya, in 1909?**

a. Theodore Roosevelt

b. Dwight Eisenhower

c. Harry Truman

d. Woodrow Wilson

Fig. 1.4.

1.8a *Only One Earth* **is a book that was published following the**

    a. UN Convention on Biological Diversity 1992

    b. UN Conference on the Human Environment 1972

    c. Paris Agreement 2015

    d. Montreal Protocol 1987

1.9a **Thomas Malthus published** *An Essay on the Principle of Population* **in**

    a. 1698

    b. 1798

    c. 1898

    d. 1998

1.10a **The Swiss chemist Paul Müller was the first person to discover the insecticidal value of**

    a. heptachlor

    b. aldrin

    c. dieldrin

    d. DDT

1.11a **Which of the following is the largest national park in the world?**

    a. Everglades National Park, Florida

    b. Serengeti National Park, Tanzania

    c. Etosha National Park, Namibia

    d. Northeast Greenland National Park, Greenland

1.12a **Match the examples of ecosystem services with the types of ecosystem services in Table 1.3.**

Table 1.3

Ecosystem service	Type of ecosystem service			
	A	**B**	**C**	**D**
Protection from flooding	Provisioning	Regulating	Provisioning	Cultural
Food from forest plants	Regulating	Provisioning	Cultural	Regulating
Spiritual experience	Cultural	Cultural	Regulating	Provisioning

    a. A

    b. B

    c. C

    d. D

1.13a Who founded the Sierra Club?

    a. John Muir

    b. Aldo Leopold

    c. John James Audubon

    d. Henry Thoreau

1.14a The concept of a biodiversity hotspot was first developed by

    a. Eugene Odum

    b. Edward Wilson

    c. Norman Myers

    d. Robert MacArthur

1.15a Which of the following organisations has provided intensive short courses in practical conservation training especially for conservation workers from developing countries?

    a. The Zoological Society of London

    b. Durrell (Jersey Zoo), Channel Islands

    c. The Bronx Zoo, New York

    d. Taronga Zoo, Australia

1.16a Tropical forests are warm but not too hot, wet but not too waterlogged and do not have a highly seasonal climate. These conditions permit high productivity, high biomass and high biodiversity. These benign characteristics of the environment are referred to as its

    a. tolerability

    b. suitability

c. fitness

d. favourableness

1.17a **Which modern-day organisation was founded in 1889 to combat the trade in plumes (feathers) for women's hats?**

a. The National Audubon Society

b. BirdLife International

c. The Royal Society for the Protection of Birds

d. The British Ornithologists' Union

1.18a **The following is a list of possible actions relating to wildlife management**

i. Take actions to make it increase

ii. Take actions to make it decrease

iii. Harvest it for a sustainable yield

iv. Leave it alone and monitor it

v. Decide on the goal of wildlife management

Which of these is included in the role of a wildlife manager?

a. i, ii and iii

b. i, iii and iv

c. i, ii, iii and iv

d. i, ii, iii, iv and v

1.19a **In many parts of the world 'wilderness' is a foreign concept. It is largely an invention of**

a. Great Britain

b. United States

c. Australia

d. Norway

1.20a Which organisation was founded in 1903 as the Society for the Preservation of the Wild Fauna of the Empire?

a. Fauna and Flora International

b. Conservation International

c. World Wildlife Fund

d. Wildlife Conservation Society

2 Environmental Pollution and Perturbations

This chapter contains questions about the pollution of air, land and water (and its effect on food chains), environmental damage and disturbance and major environmental disasters.

Foundation

2.1f When organic matter of human origin enters a lake the water may quickly become enriched and suffer serious damage due to eutrophication. This process is usually referred to as

a. traditional eutrophication

b. cultural eutrophication

c. organic eutrophication

d. enriched eutrophication

2.2f Tropical deforestation is often accomplished by a method of cutting vegetation known as

a. cut-and-burn

b. hack-and-burn

c. slash-and-burn

d. slash-and-blaze

2.3f The source of most of the plastic found in the oceans is material that enters marine water from

   a. beaches

   b. rivers

   c. coastal industries

   d. ships

2.4f A fast breeder nuclear reactor generates

   a. more fissile material than it consumes

   b. the same amount of fissile material as it consumes

   c. no fissile material

   d. less fissile material than it consumes

2.5f Which of the following will not cause eutrophication if released into a lake?

   a. Cadmium waste

   b. Chemical fertiliser

   c. Milk

   d. Sewage

2.6f Acid rain has a pH of

   a. <3.5

   b. <4.5

   c. <5.6

   d. <7.0

2.7f Chemicals that were widely used in the past as a propellant in aerosol cans, foam packaging, refrigerators, air conditioning systems and solvents caused damage to which layer of the atmosphere?

   a. The ozone layer

   b. The thermosphere

   c. The troposphere

   d. The stratosphere

2.8f   Which of the following has/have been important sources of lead contamination in the environment in the past?

a. Paint

b. Gunshot pellets

c. Petrol

d. All of the above

2.9f   Plants able to grow successfully on soil contaminated by heavy metals are known as

a. heavy metal resistant

b. heavy metal tolerant

c. heavy metal selected

d. heavy metal lenient

2.10f   Which of the following was not a source of a major marine oil spill?

a. *Torrey Canyon*

b. *Exxon Valdez*

c. *Amoco Cadiz*

d. *Herald of Free Enterprise*

2.11f   Heavy metals are important environmental pollutants that are, for example, found in mine tailing and sewage sludge, and released into water and air from industrial processes. Which of the following lists contains an element that is not a heavy metal?

a. Tin, potassium, silver

b. Mercury, selenium, zinc

c. Cadmium, lead, titanium

d. Copper, iron, nickel

2.12f   Which of the following signs is typical of advanced mercury poisoning in adult humans?

a. Lack of coordination and difficulty walking

b. Muscle twitching

21

c. Problems with hearing, speech and vision

d. All of the above

2.13f **The phenomenon whereby some pollutants are concentrated in organisms as they are passed from one organism to another in a food chain is called**

a. bioaddition

b. biomultiplication

c. bioaccumulation

d. bioduplication

2.14f **The main interface between air pollution and the interior of a flowering plant is provided by**

a. the stomata

b. the parenchyma

c. the collenchyma

d. the sclerenchyma

2.15f **Which of the following sources of pollution is not a point source?**

a. Noise pollution from a jet flying overhead

b. Water pollution from a factory entering a river from a drain pipe

c. Air pollution from a forest fire

d. Light pollution from a street light

2.16f **Between 1980 and 2000, 100 million hectares of tropical forest were lost, mostly due to**

a. palm oil plantations in South America and cattle ranching in Southeast Asia

b. cattle ranching in South America and palm oil plantations in Southeast Asia

c. crop production in South America and palm oil plantations in Southeast Asia

d. coffee production in West and East Africa and Cattle ranching in South America

2.17f   Large quantities of which of the following gases are released in leaks from wells created by hydraulic fracturing (fracking)?

    a. Nitrogen and carbon monoxide

    b. Hydrogen and methane

    c. Methane and carbon dioxide

    d. Carbon dioxide and ozone

2.18f   Polychlorinated biphenyls (PCBs) are

    a. persistent organic pollutants

    b. inorganic acids

    c. difficult to destroy by incineration

    d. unable to pass through the human placenta

2.19f   Between 1700 and 2000, which of the following habitats lost the highest proportion (13%) of its global area?

    a. Tropical forests

    b. Temperate forests

    c. Wetlands

    d. Savannahs

2.20f   Which of the following is a secondary pollutant?

    a. Carbon monoxide

    b. Sulphur dioxide

    c. Mercury

    d. Ozone

Intermediate

2.1i   The 'feminisation' of individuals in some populations of fishes and amphibians in some rivers is caused by

    a. farm animal slurry containing antibiotics and growth hormones

    b. contaminated water from nuclear facilities

c. chemicals discharged into rivers from the pulp and paper industry

d. hormones from human contraceptive pills discharged from wastewater treatment plants

2.2i **Which of the following was not the site of a major nuclear incident?**

a. Bhopal, India

b. Three Mile Island, United States

c. Chernobyl, former USSR

d. Windscale/Sellafield, United Kingdom

2.3i **Which of the following activities is a cause of soil erosion?**

a. Overgrazing by cattle in the southwest United States

b. The clearance of land for crops in Brazil

c. The removal of trees in the Sahel

d. All of the above

2.4i **Forests in which of the following areas have suffered most from the effects of acid rain?**

a. Spain and Portugal

b. Sweden and Norway

c. Germany and Poland

d. Hungary and Romania

2.5i **Minamata disease is named after a village in Japan where the human population suffered deaths and serious medical conditions caused by contamination of the marine food chain with**

a. zinc

b. lead

c. mercury

d. cadmium

2.6i   Which of the following water bodies has dried up as a result of water being diverted to irrigate cotton crops?

    a. The Black Sea

    b. The Aral Sea

    c. The Adriatic Sea

    d. The Baltic Sea

2.7i   The Three Gorges Dam Project displaced a large number of people and flooded an extensive area of land in

    a. Vietnam

    b. Myanmar

    c. China

    d. India

2.8i   Which of the following pollutants causes respiratory difficulties by combining with haemoglobin?

    a. Sulphur dioxide

    b. Carbon monoxide

    c. Nitrogen oxide

    d. Ozone

2.9i   If mercury is released into marine waters which of the following groups of organisms would you expect to accumulate the highest concentrations?

    a. Predatory fishes

    b. Zooplankton

    c. Phytoplankton

    d. Herbivorous fishes

2.10i Which of the following statements about rivers and river catchments is false?

    a. Sediment discharge increases when a catchment is deforested

    b. Total sediment yields are higher in urban areas than rural areas of similar size

c. During summer, storms are likely to add heat to rivers in urban areas

d. Nutrient cycles are not disrupted when catchments are deforested

2.11i **Which of the following statements about the effects of oil pollution of marine waters is false?**

a. Oiled seabirds may die of hypothermia because oil interferes with the insulating properties of their feathers

b. The animal and plant populations on oiled beaches recover faster if detergent is applied to the oil than if it is not

c. Oiled seabirds are frequently poisoned by oil ingested while preening in an attempt to clean themselves

d. In response to some oil pollution incidents oil has been sunk by covering it with chalk and soil

2.12i **The quantity of strontium-90 in cows' milk in Britain increased manyfold in 1964 as a result of**

a. volcanic eruptions

b. the testing of nuclear weapons

c. an explosion at a large nuclear power plant

d. accidental releases of nuclear waste

2.13i **Which of the following interferes with the navigational abilities of cetaceans?**

a. Sonar used by military submarines

b. Signals from communications satellites

c. Microwaves from mobile phone masts in coastal areas

d. Bright lights from coastal towns

2.14i **The effects of DDT on birds of prey populations in Britain have been examined using museum specimens of sparrowhawks (*Accipiter nisus*). Fig. 2.1 shows a steep drop between 1940 and 1950 in the value of variable *y* which represents**

a. DDT concentration in feathers

b. egg shell thickness

c. adult body size in females

d. DDT deposits in fatty tissues

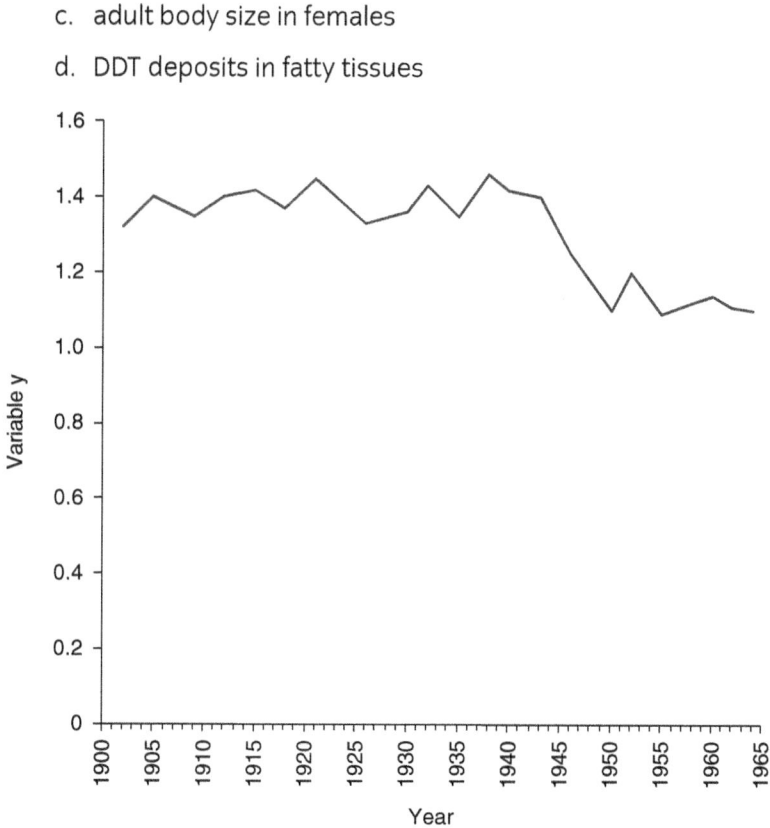

Fig. 2.1.

2.15i **Dioxins are persistent, highly toxic environmental pollutants that, in animals, mainly accumulate in the**

a. kidneys

b. brain

c. muscles

d. fatty tissue

2.16i **Which of the following isotopes has the longest half-life?**

a. Iodine-131

b. Plutonium-239

c. Strontium-90

d. Uranium-238

2.17i The accident at the Chernobyl nuclear power station in Ukraine in 1986 released strontium-90 (^{90}Sr) that contaminated large areas of land including hill pasture ecosystems in Britain. Which component of these ecosystems would you have expected to contain the highest concentrations of ^{90}Sr?

a. Trees

b. Soil

c. Sheep

d. Grass

2.18i Catalytic converters are responsible for increases in the environment of which of the following metals?

a. Platinum, palladium and rhodium

b. Silver, lead and platinum

c. Cadmium, chromium and thallium

d. Titanium, copper and rhodium

2.19i A PM$_{2.5}$ Health Advisory is a warning of high levels of which pollutants in outdoor air?

a. Volatile organic compounds

b. Nitrogen oxides

c. Carbon monoxide

d. Very small particles

2.20i The water body in Fig. 2.2 is partially covered in small aquatic plants known as duckweed. This is an indication that the water is

a. eutrophic

b. anoxic

c. oligotrophic

d. toxic

Fig. 2.2.

Advanced

2.1a   In the ocean, microplastics are defined as pieces of plastic that are smaller than

a. 0.5mm

b. 1.0mm

c. 5.0mm

d. 10.0mm

2.2a   The graph below (Fig. 2.3) allows us to visualise how well-adapted an organism is to its environment and is a simple representation of a

a. fitness landscape

b. suitability landscape

c. fitness setting

d. survival landscape

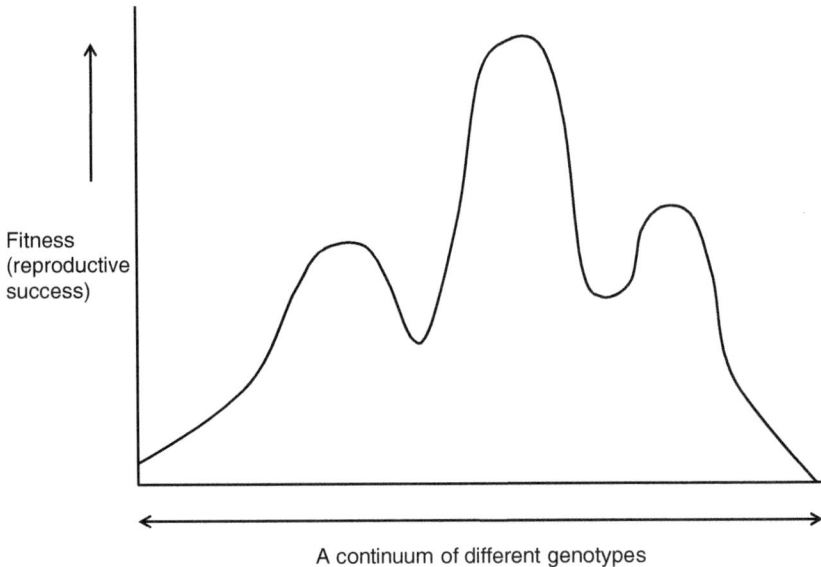

A continuum of different genotypes

Fig. 2.3.

2.3a   Transuranic waste is a type of

   a. nuclear waste

   b. plastic waste

   c. organic waste

   d. mine waste

2.4a   The process known as 'coral bleaching' involves

   a. the loss of medusae from corals

   b. the loss of calcium from corals

   c. the accumulation of sediment on corals

   d. the loss of zooxanthellae from corals

2.5a   The decline of populations of vultures (*Gyps* spp.) in Asia has been attributed to

   a. the use of diclofenac, an anti-inflammatory drug given to cattle

   b. the use of antibiotics to treat livestock

   c. the use of vulture bones in traditional medicine

   d. deliberate poisoning

2.6a **Which of the graphs in Fig. 2.4 shows a temperature inversion?**

a. A

b. B

c. C

d. D

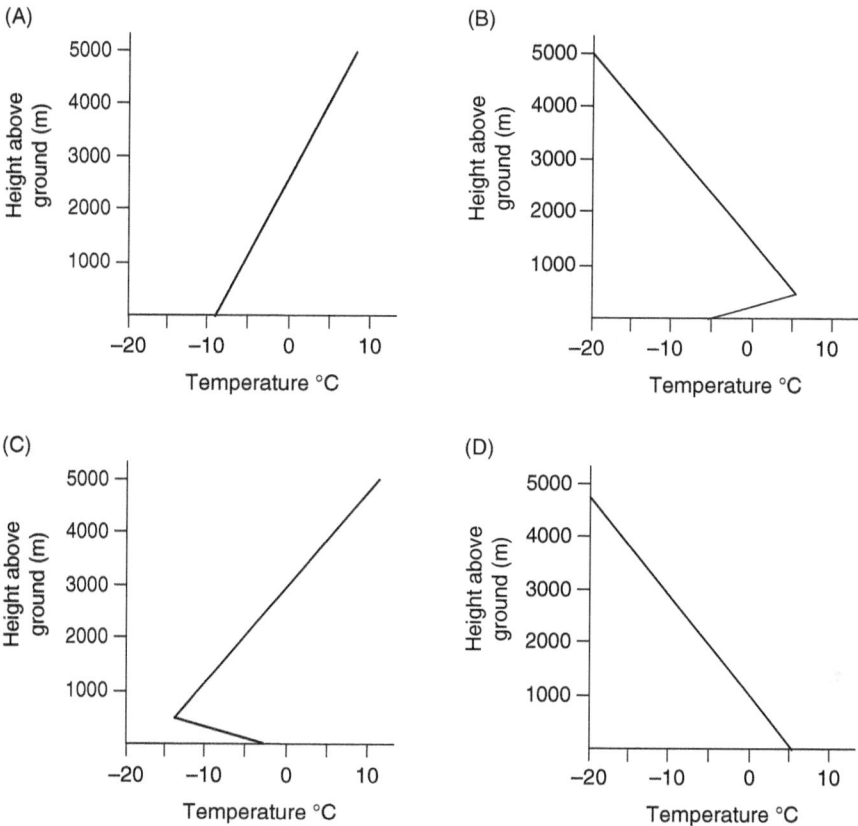

(A)

(B)

(C)

(D)

Fig. 2.4.

2.7a **Love Canal is a location in New York where**

a. a nuclear power station caused a release of radioactive waste

b. a watercourse was polluted by drainage water from an abandoned coal mine

c. dumped toxic chemicals caused birth defects and disease in the local people

d. a spill from an oil tanker caused extensive damage to local wildlife

2.8a Neonicotinoids are

    a. herbicides that are widely used in agriculture to kill weeds

    b. insecticides that have been implicated in the decline of some insect pollinators

    c. pheromones used in agriculture to attract insect pollinators

    d. human hormones that are excreted in urine and enter rivers where they reduce freshwater insect diversity

2.9a **Which of the following pairs of abbreviations are both breakdown products of the pesticide DDT?**

    a. DDE and DDD

    b. DDC and DDE

    c. DDD and DDC

    d. DDF and DDD

2.10a **The diagram below (Fig.2.5) shows the chemical structure of**

    a. lindane

    b. DDT

    c. 2,4-D

    d. dieldrin

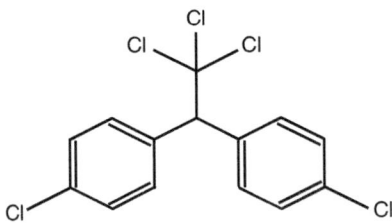

Fig. 2.5.

2.11a **Which of the following chemicals was largely responsible for creating the 'hole' in the ozone layer?**

    a. Polychlorinated biphenyls

    b. Chlorofluorocarbons

c. Dichlorodiphenyltrichloroethane

d. Sulphur dioxide

2.12a Acid rain causes a particular problem for aquatic molluscs because

a. it causes eutrophication

b. it interferes with osmoregulation

c. it breaks down the calcium carbonate they need to make their shells

d. it reacts with oxygen making it difficult for them to obtain sufficient oxygen for respiration

2.13a Which of the following statements about DDT is false?

a. It is likely to be present in higher concentrations in top predators than in herbivores

b. It does not travel in the atmosphere

c. It accumulates in fatty tissues

d. It was once widely used in tropical areas in an attempt to reduce the incidence and spread of malaria

2.14a Which of the following statements describes a temperature inversion and its effects?

a. Cold air becomes trapped under warm air, for example, on a hot day following a cold night, increasing the dispersal of atmospheric pollutants

b. Warm air becomes trapped under cold air, for example, on a cold day following a warm night, reducing the dispersal of atmospheric pollutants

c. Cold air becomes trapped under warm air, for example, on a hot day following a cold night, reducing the dispersal of atmospheric pollutants

d. Warm air becomes trapped under cold air, for example, on a cold day following a warm night, increasing the dispersal of atmospheric pollutants

2.15a **A chemical that can cause birth defects is referred to as being**

    a. teratogenic

    b. carcinogenic

    c. mutagenic

    d. ontogenetic

2.16a **A nonthreshold pollutant is one in which**

    a. there is a finite dose below which adverse health effects are not discernable

    b. there is some health risk at any level of exposure

    c. there is no health risk at any level of exposure

    d. the threshold level beyond which health risks are detectable has not been determined

2.17a **An air pollution emission plume that is warmer and lighter than the ambient air is known as a**

    a. dense gas plume

    b. passive plume

    c. neutral plume

    d. buoyant plume

2.18a **The greatest natural disasters (perturbations) on Earth are caused by**

    a. impacts with celestial bodies

    b. floods

    c. forest fires

    d. earthquakes

2.19a **Peroxyacyl nitrates (PANs) are present in**

    a. inorganic fertilisers

    b. untreated sewage

    c. photochemical smog

    d. mine waste

2.20a The 'Baby Tooth Survey' was conducted in the United States in the 1960s and examined the deciduous teeth of children to investigate the effects of

a. lead pollution in the air

b. mercury in the diet

c. DDT in food crops

d. nuclear fallout from atomic bomb tests

3 Wildlife and Conservation Biology

This chapter contains questions about wildlife management and conservation, endangered species, nature reserve design and the role of zoos in conservation.

Foundation

3.1f **Which of the following has not been saved from extinction by captive breeding?**

    a. Père David's deer (*Elaphurus davidianus*)

    b. Black-footed ferret (*Mustela nigripes*)

    c. California condor (*Gymnogyps californianus*)

    d. Mountain gorilla (*Gorilla beringei beringei*)

3.2f **A species whose protection indirectly safeguards many other species in the ecological community is known as**

    a. a flagship species

    b. an indicator species

    c. an umbrella species

    d. a pioneer species

3.3f **The crown of thorns starfish (*Acanthaster planci*) is infamous because it has caused extensive damage to**

    a. the Great Barrier Reef, off the east coast of Australia

    b. the Galapagos Islands, off the west coast of Ecuador

c. marine areas around the Hawaiian Islands

d. coastal ecosystems in the Gulf of Mexico

3.4f   A strip of protected land linking two larger protected areas is called a

a. protected path

b. wildlife access strip

c. wildlife corridor

d. wildlife route

3.5f   The survival of orangutans (*Pongo* spp.) in the wild is threatened by the expansion of plantations of

a. coffee

b. palm oil

c. tea

d. cotton

3.6f   Which of the following is a non-consumptive use of nature?

a. Ecotourism

b. Commercial hunting

c. Fuelwood cutting

d. Commercial fishing

3.7f   Conspicuous and attractive species used to publicise or raise funds for a habitat or area of conservation interest is called a

a. pennant species

b. banner species

c. poster species

d. flagship species

3.8f   Which of the following groups exhibits a decrease in species diversity from the poles to the tropics?

a. Marine fishes

b. Terrestrial mammals

c. Conifers

d. Birds

3.9f **The cooperative breeding programme known as the Species Survival Plan Program is operated by**

a. the European Association of Zoos and Aquaria (EAZA)

b. the Association of Zoos and Aquariums (AZA)

c. the British and Irish Association of Zoos and Aquariums (BIAZA)

d. the Southeast Asian Zoos and Aquariums Association (SEAZA)

3.10f **The terms most likely to be used for places where animal gametes and plant propagules are stored in very cold conditions are**

a. frozen zoo and plant bank

b. seed bank and frozen zoo

c. gamete bank and propagule bank

d. seed store and cool zoo

3.11f **The creation of protected areas for wildlife involving the displacement of the indigenous people, seizure of their land and the curtailing of their customary rights to water, hunting, fishing and other resources is called**

a. stronghold conservation

b. segregation conservation

c. fortress conservation

d. citadel conservation

3.12f **Red foxes (*Vulpes vulpes*) are not native to Australia but were taken there by the British so that the colonists could continue to engage in the traditional sport of fox hunting. This release of red foxes is most accurately described as**

a. a reintroduction

b. population supplementation

c. a translocation

d. an introduction

3.13f A conservation project for chimpanzees (*Pan troglodytes*) may not be classified as *ex-situ* if it occurs in

a. a South African zoo

b. London Zoo

c. the wild in a Tanzanian National Park

d. a sanctuary in the United Kingdom

3.14f Which of the following methods has/have been used to manage elephant (*Loxodonta africana*) populations in Africa?

a. Culling

b. Contraception

c. Translocation

d. All of the above

3.15f Major routes along which migratory birds fly from their over-wintering grounds to their breeding grounds are called

a. flyways

b. migration paths

c. birdways

d. skyways

3.16f Which of the following pairs of protected areas is important in protecting the wildebeest (*Connochaetes taurinus*) that migrate annually between Kenya and Tanzania?

a. Tsavo East National Park and the Serengeti National Park

b. Selous Game Reserve and Nairobi National Park

c. Amboseli National Park and Etosha National Park

d. Serengeti National Park and the Maasai Mara National Reserve

3.17f A population of a rare species held and bred in captivity for conservation purposes is often referred to as

a. a safeguarded population

b. a rescue population

c. a guarantee population

d. an insurance population

3.18f **A great deal of conservation effort has been focussed on protecting the 'megafauna'. This term refers primarily to**

   a. mammals of large body size

   b. animals of large body size

   c. mammals with extensive ranges

   d. animals with extensive ranges

3.19f **The document used by zoos to manage a cooperative breeding programme for animals is called**

   a. a studbook

   b. a database

   c. a pedigree

   d. a genealogy

3.20f **Which of the following could not be an example of *in-situ* conservation?**

   a. A human–elephant conflict project in Assam, India, aimed at reducing elephant deaths by monitoring their movements

   b. A captive breeding programme for black rhinoceroses conducted in European zoos

   c. The translocation of wild golden eagles from Scotland to Ireland

   d. The creation of new ponds on farmland in England

Intermediate

3.1i **The Species Survival Commission (SSC) is part of the**

   a. International Union for the Conservation of Nature (IUCN)

   b. World Wide Fund for Nature (WWF)

   c. World Association of Zoos and Aquariums (WAZA)

   d. Zoological Society of London (ZSL)

3.2i   A population of animals that is destined to become extinct by virtue of having insufficient reproductive capacity to recover its numbers due to low fecundity, poor offspring survival, or for some other reason, is referred to as being

   a. ecologically extinct

   b. virtually extinct

   c. demographically extinct

   d. biologically extinct

3.3i   In 2020, how many areas of the world were classified as biodiversity hotspots by Conservation International?

   a. 16

   b. 26

   c. 36

   d. 46

3.4i   Which of the following areas is not considered to be a biodiversity hotspot?

   a. Madagascar

   b. The Western Ghats, India

   c. The Atlantic Forest, Brazil

   d. Tasmania, Australia

3.5i   The search for, and exploitation of, components of the biodiversity of a country for drug development or some other commercial use is called

   a. bioutilisation

   b. bioprospecting

   c. bioexploitation

   d. biomonetisation

3.6i   Which area of ecology has been particularly important in the theoretical study of the design of nature reserves?

   a. Population dynamics

   b. Island biogeography

      c. Production ecology

      d. Ecological genetics

3.7i **Which animals have been responsible for the extinction of several species of marsupials in Australia, the extirpation of seabirds on oceanic islands and the predation of tortoises on the Galapagos Islands?**

      a. Cats

      b. Dogs

      c. Foxes

      d. Mink

3.8i **Species that are susceptible to extinction by virtue of having a limited capacity to recover when their populations are heavily depleted (e.g. small litter or clutch size, long generation time) are referred to as being**

      a. *t*-selected species

      b. *r*-selected species

      c. *K*-selected species

      d. *N*-selected species

3.9i **Protected areas known as 'peace parks', because they foster cooperation and political goodwill, are located in**

      a. international waters

      b. transboundary areas

      c. areas of great human–animal conflict

      d. war zones where wildlife is threatened

3.10i **Habitats and species can be maintained if they can be shown to provide direct value to the local people with whom they coexist. This approach to conservation is known as the**

      a. 'exploit it and value it' approach

      b. 'exploit it and save it' approach

c. 'value it or lose it' approach

d. 'use it or lose it' approach

3.11i **Red, amber and green lists are used to categorise the birds in the United Kingdom by conservation status by the**

a. IUCN

b. RSPB

c. WWF

d. FFI

3.12i **Which of the following groups of IUCN categories are classified together as 'threatened'?**

a. Critically Endangered (CR) and Endangered (EN) only

b. Critically Endangered (CR), Endangered (EN) and Vulnerable (VU)

c. Endangered (EN) and Vulnerable (VU) only

d. Vulnerable (VU) and Critically Endangered (CR) only

3.13i **The concept of biodiversity hotspots has been criticised by some ecologists because it**

a. risks creating a world where a relatively small proportion of the Earth is preserved with high biodiversity while other areas are allowed to deteriorate

b. causes funding to be directed towards biodiversity instead of economic development

c. focuses too much on the richness of animal species

d. focuses too much on tropical ecosystems

3.14i **Which of the following is not a goal of all captive breeding programmes?**

a. Maintenance of a healthy age structure

b. Protection of the population against disease

c. Preservation of the gene pool to avoid inbreeding

d. Release back to the wild

3.15i **Conservationists managing captive breeding programmes attempt to avoid genetic problems in future generations of rare species by calculating**

    a. conservation coefficients

    b. correlation coefficients

    c. inbreeding coefficients

    d. genetic coefficients

3.16i **Which of the following statements about rhinoceros horn is false?**

    a. Scientists have developed fake rhino horn made of horse hair bound with a protein cement that could destabilise the market for real rhino horn because buyers would not be able to distinguish between the two

    b. Rhino horn is an aphrodisiac

    c. Rhino horn is used in traditional Chinese medicine to treat fever

    d. A main driver of the illegal trade in rhino horn is the need of some rich people, predominantly in Asia, to exhibit their wealth by owning it

3.17i **Which of the following sequences of IUCN categories indicates a progressively increasing risk of extinction?**

    a. Near Threatened, Vulnerable, Endangered

    b. Vulnerable, Endangered, Near Threatened

    c. Endangered, Critically Endangered, Vulnerable

    d. Endangered, Vulnerable, Extinct in the Wild

3.18i **What percentage of all the animal species that have become extinct since 1600 were on islands?**

    a. 25%

    b. 50%

    c. 75%

    d. 90%

3.19i **Which of the following was the world's most trafficked wild mammal in 2020?**

    a. Sunda pangolin (*Manis javanica*)

    b. Sumatran tiger (*Panthera tigris sumatrae*)

    c. Saiga (*Saiga tatarica*)

    d. Hawksbill turtle (*Eretmochelys imbricata*)

3.20i **The structure in Fig.3.1 is used to protect crops in parts of Africa and Asia by deterring wild elephants thereby reducing human–elephant conflict. The wooden boxes contain**

    a. wasps

    b. pepper

    c. ants

    d. bees

Fig. 3.1.

Advanced

3.1a The minimum population size required to provide some specified probability that the population will survive for a given period of time is known as the

    a. minimum viable density

    b. minimum reproductive population

    c. effective population size

    d. minimum viable population

3.2a A demographically isolated population whose probability of extinction over the time scale of interest (perhaps 100 years) is not substantially affected by natural immigration from other populations is called an

    a. ECU

    b. ESS

    c. ESU

    d. ECS

3.3a Animals of the same species held in several zoos and exchanged between them for captive breeding purposes comprise a

    a. subpopulation

    b. megapopulation

    c. metapopulation

    d. micropopulation

3.4a The arrival of immigrants to a sink patch may top up a population so that it does not become extinct due to an inability to produce sufficient new individuals by reproduction. This is known as the

    a. recovery effect

    b. rescue effect

    c. salvation effect

    d. migration effect

3.5a Regional populations of a species may be established over great distance and barriers by a process called

a. leap dispersal

b. hop dispersal

c. spring dispersal

d. jump dispersal

3.6a Which of each of the pairs of diagrams in Fig.3.2 (A or B and C or D) illustrate the best arrangements of nature reserves (indicated by grey circles) based on the principles described by J.M. Diamond (1975)?

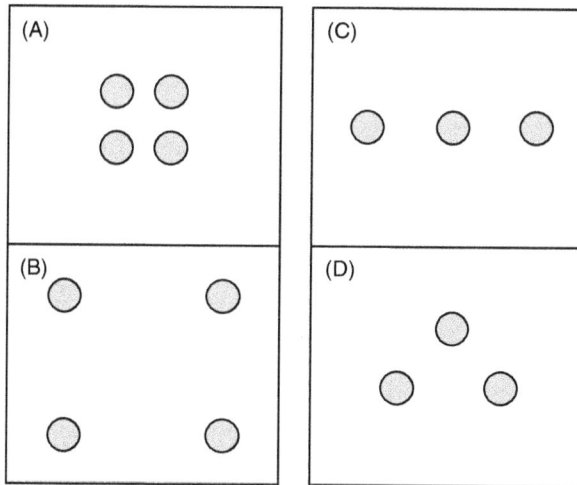

Fig. 3.2.

a. A and C

b. A and D

c. B and C

d. B and D

3.7a Birdlife International defines 'Restricted Range Birds' (RRBs) as species whose global distribution is less than 50,000 km². An 'Endemic Bird Area' (EBA) is a site with two or more RRBs. These tend to be clustered on

a. islands in the tropics

b. mountains in the tropics

c. islands and mountains in the tropics

d. islands and mountains in temperate latitudes

3.8a The dates of mass extinctions of bird species on Pacific islands are closely linked to

a. the first arrival of humans

b. the arrival of invasive species

c. outbreaks of disease

d. climate change

3.9a The valuation of biodiversity based on the amount of money people say they are prepared to pay, or forgo, to fund conservation is called its

a. beneficial valuation

b. ecological valuation

c. contingent valuation

d. bequest valuation

3.10a Species A is divided into two subspecies A1 and A2. The subspecies are treated as separate ESUs for captive breeding purposes. Some A1 x A2 hybrids exist in the captive population from a time when the two subspecies had not been identified. These hybrids have not been used for further breeding. New scientific work suggests that subspecies A2 could be divided into two so that the species would exist as three subspecies A1, A2 and A3. The global captive population of species A is very low. Should conservationists support the acceptance of subspecies A3 as a new ESU?

a. No, because there are already A1 x A2 hybrids in the captive population and this would mean testing A2 x A2 individuals to see if any are in fact A2 x A3 hybrids and useless for further breeding if A3 is accepted as a new ESU

b. Yes, because we need to conserve as much biodiversity as possible

c. This decision should be left up to individual zoos

d. The situation described here is fictitious and unlikely ever to occur in reality

3.11a The biodiversity hotspots identified by Norman Myers in 2000 were

    a. regions that contain large numbers of charismatic species

    b. regions with high concentrations of endemic animal species that have suffered severe habitat destruction

    c. regions with high concentrations of endemic plant species that have suffered severe habitat destruction

    d. regions with high species richness

3.12a The theoretical interplay between the immigration of species to, and extinction of species on, an island is illustrated in Fig. 3.3. Which of the following is true for the lines labelled A and C and the point labelled B?

    a. A is immigration rate; B is equilibrium richness; C is extinction rate

    b. A is extinction rate; B is equilibrium richness; C is immigration rate

    c. A is equilibrium rate; B is immigration rate; C is extinction rate

    d. A is immigration rate; B is extinction rate; C is equilibrium richness

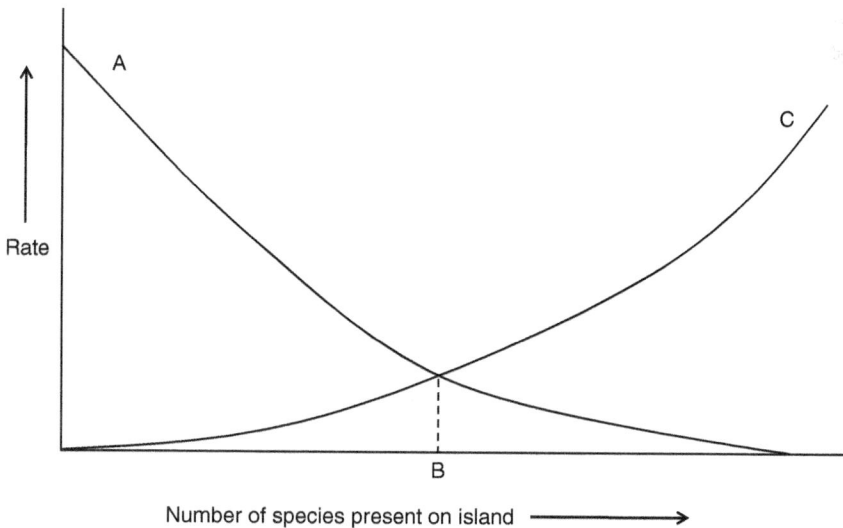

Fig. 3.3.

3.13a CRISPR is a technology used to

    a. create databases of endangered species

    b. manage inbreeding in zoo populations of rare species

    c. model changes in animal populations

    d. edit genomes

3.14a The EDGE of Existence Programme of the Zoological Society of London is concerned with the conservation of species that are

    a. Ecologically Depleted and Globally Endangered

    b. Evolutionarily Distinct and Greatly Endangered

    c. Ecologically Distinct and Globally Endangered

    d. Evolutionarily Distinct and Globally Endangered

3.15a The circles in Fig. 3.4 represent different genotypes of the same species. The diagram is a representation of changes in the proportions of different genotypes with time and shows a

    a. genetic obstacle

    b. genetic obstruction

    c. genetic bottleneck

    d. genetic blockage

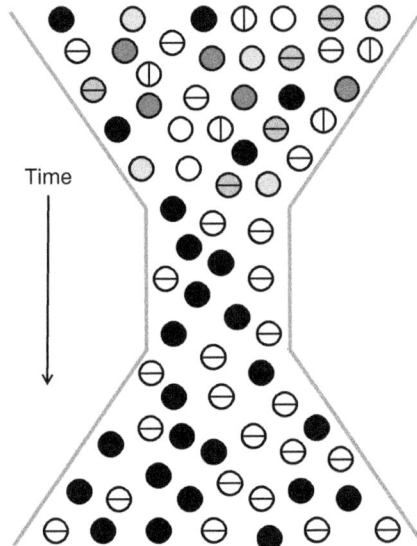

Fig. 3.4.

3.16a **The objects in Fig. 3.5 are hanging from power lines. They are made from reflective materials and spin around their vertical axis in the wind. Near which of the following are they most likely to be located?**

   a. A nature reserve near an airport

   b. A nature reserve visited by large numbers of wildfowl

   c. Agricultural land growing cereal crops

   d. An electricity-generating facility

Fig. 3.5.

3.17a **A foal of which of the following equine taxa was born to a surrogate mother in a veterinary facility in Texas in August 2020 as a result of the first successful cloning of the species using DNA that had been cryopreserved?**

   a. Mountain zebra (*Equus zebra*)

   b. Somali wild ass (*Equus africanus somaliensis*)

   c. Przewalski's horse (*Equus ferus przewalskii*)

   d. Onager (*Equus hemionus*)

3.18a Some mountain lions (*Puma concolor*) in the Santa Monica Mountains in Los Angeles, California, have L-shaped kinks at the end of their tail. This population is isolated from others by urban development and highways. Wildlife biologists believe this is the result of

    a. inbreeding

    b. collisions with vehicles

    c. mutation

    d. interspecific aggression

3.19a The Clovis overkill extinctions occurred between 11,000 and 15,000 years ago in

    a. North and South America

    b. Africa

    c. Europe

    d. Australia

3.20a When human activity encroaches into natural areas human–wildlife conflict often occurs. During the construction of the Kenya–Uganda railway in 1898 many workers (possibly as many as 135) were killed by two big cats that came to be known as the 'Maneaters of Tsavo'. They were eventually shot and their stuffed bodies were subsequently displayed in the Field Museum of Natural History in Chicago. They were both

    a. female leopards

    b. male leopards

    c. female lions

    d. male lions

4 Restoration Biology and Habitat Management

This chapter contains questions about habitat management, the restoration of lost and damaged habitats, captive breeding and the reintroduction of species to their natural habitats.

Foundation

4.1f The individual organisms used to initiate a captive breeding programme are called the

   a. initiators

   b. originators

   c. founders

   d. establishers

4.2f Grey wolves (*Canis lupus*) have re-established populations in Western Europe by natural

   a. dispersion

   b. dispersal

   c. diffusion

   d. dissemination

4.3f A species whose activities substantially modify the structure of the environment, such as the beaver (*Castor fiber*), is called an

   a. ecosystem engineer

   b. ecosystem architect

c. ecosystem modifier

d. ecosystem developer

4.4f **Which country constructed fences to control the movement of rabbits (*Oryctolagus cuniculus*) in the 1880s?**

a. South Africa

b. United States

c. Australia

d. Scotland

4.5f **Which of the following ecosystems can be restored by organisms colonising and growing on metal frames?**

a. Sand dunes

b. Marshlands

c. Coral reefs

d. Mangroves

4.6f **The Earth's stocks of natural assets, including animals, plants, air, water, soil and geology make up its**

a. ecosystem services

b. ecological resilience

c. ecological capital

d. natural capital

4.7f **The establishment of forest growth on an area that was not previously covered by forest is most accurately known as**

a. forestation

b. deforestation

c. afforestation

d. reforestation

4.8f **Which of the following is a restoration technique that is used to quickly re-establish plants on steep slopes of bare soil?**

a. Hydroseeding

b. Broadcast seeding

  c. Seed drilling

  d. Dibbling

4.9f   **In 2005 the Kenya Wildlife Service (KWS) moved approximately 150 elephants (*Loxodonta africana*) from Shimba Hills National Reserve to Tsavo East National Park. This event is most accurately described as**

  a. a reintroduction

  b. an introduction

  c. a resettlement

  d. a translocation

4.10f   **The British evolutionary ecologist Prof. A. D. Bradshaw used the terms 'restoration', 'replacement' and 'rehabilitation' to describe different outcomes of restoration projects (Bradshaw, 1987). Match the definitions to the correct list of terms in Table 4.1.**

Table 4.1

Definition	A	B	C	D
The original ecosystem is produced	Replacement	Restoration	Restoration	Replacement
The original ecosystem is partially produced	Rehabilitation	Replacement	Rehabilitation	Restoration
An alternative ecosystem is established	Restoration	Rehabilitation	Replacement	Rehabilitation

  a. A

  b. B

  c. C

  d. D

4.11f A captive-bred European wildcat (*Felis silvestris silvestris*) is contained within a holding pen at a release site to monitor its progress prior to release to the wild. This is known as a

 a. hard release

 b. soft release

 c. staged release

 d. delayed release

4.12f Restoring an extensive area of salt marsh can help to protect a coastline and human settlements from damage during severe storms. This habitat can be considered to be providing an

 a. ecosystem service

 b. ecosystem benefit

 c. ecosystem protection

 d. ecosystem barrier

4.13f 'Operation Oryx' was a project whose aim was to reintroduce the Arabian oryx (*Oryx leucoryx*) to parts of the Middle East. Which of the following statements about this project is false?

 a. It involved San Diego Zoo in California

 b. It was a funded by the World Wildlife Fund (WWF)

 c. Some oryx were returned to Jordan

 d. Some oryx were returned to Oman

4.14f Rows of trees planted to prevent the spread of a desert resulting from wind-blown sand are referred to as a

 a. tree screen

 b. sand shield

 c. windbreak

 d. shelterbelt

4.15f **When a cleanup of chemical contaminants is undertaken on an old industrial site prior to the land being repurposed the process is called**

    a. remediation

    b. renovation

    c. rectification

    d. rejuvenation

4.16f **When a second runway was constructed at Manchester International Airport in the United Kingdom an area of woodland was planted that was six times the area lost during construction, a wildflower grassland was created, 30,000 newts, frogs and toads were translocated to over 90 new or restored ponds, new barns were constructed as bat roosts and a family of badgers (*Meles meles*) was relocated to an artificial sett. This process of compensating for the loss of habitat is referred to as**

    a. alleviation

    b. mitigation

    c. reparation

    d. compensation

4.17f **When the original forest that grew in an area is felled the forest that replaces it is called a**

    a. replacement forest

    b. substitute forest

    c. derived forest

    d. secondary forest

4.18f **Some marine areas at risk of overfishing have been protected by the establishment of Marine Protected Areas (MPAs) in an attempt to promote the recovery of fish populations. When fishing is banned in an MPA studies have shown that**

    a. fish biomass increases in the MPA but decreases in the adjacent unprotected (fished) area

    b. fish biomass increases in the MPA but remains unchanged in the adjacent unprotected (fished) area

c. fish biomass increases in the MPA and increases in the adjacent unprotected (fished) area

d. fish biomass remains unchanged in the MPA but increases in the adjacent unprotected (fished) area

4.19f In the early stages of a restoration project native plants may be outcompeted by invasive species that grow faster. The native species may be helped by inoculation with

a. mycorrhizal fungi

b. antibiotics

c. fertilsers

d. nematodes

4.20f The removal of heavy metals from water systems using biological materials is known as

a. bioleaching

b. biosynthesis

c. bioretention

d. biosorption

Intermediate

4.1i A conservation programme that involves young animals, that have been captive-bred or taken from the wild, being reared in captivity until they reach a larger size and then releasing them to the wild is known as

a. an advancement programme

b. as assistance programme

c. a headstarting programme

d. a facilitation programme

4.2i Meadows in Europe are mown to provide food and litter for livestock. Delaying the first mowing from spring to summer

a. increases plant and invertebrate species richness

b. decreases plant and invertebrate species richness

c. has no effect on plant and invertebrate species richness

d. increases plant species richness only

4.3i **A managed area of wetland covered in sphagnum moss and set aside specifically for the purpose of removing carbon from the atmosphere to offset carbon dioxide emissions is called a**

a. conversion farm

b. carbon farm

c. conservation farm

d. carbon dioxide farm

4.4i **An ecoduct is an alternative name for**

a. a wildlife bridge or tunnel constructed to allow animals to cross barriers such as roads

b. a channel that allows fish to bypass a weir in a river

c. a pipe that carries treated sewage from a wastewater treatment plant to a river

d. a type of aqueduct that carries a canal over a valley

4.5i **Moorlands in northern England are managed for red grouse (*Lagopus lagopus scotica*) shooting by**

a. removing hedgerows

b. draining peat bogs

c. burning heather

d. tree planting

4.6i **If the construction of a new highway results in the destruction of a large area of ancient forest and a new 'forest' is planted by the construction company as a mitigation measure to 'replace' this, the new forest is most accurately described as**

a. an afforestation project

b. a reforestation project

c. a forestation project

d. a deforestation project

4.7i **If habitat components (e.g. soil and trees) are moved from one site (the 'donor site') to another (the 'receptor' site), this process is called**

    a. habitat translocation

    b. habitat translocation, but only if the receptor site formerly supported the newly create habitat

    c. habitat introduction, but only if the receptor site formerly supported the newly create habitat

    d. habitat reintroduction

4.8i **Oregon Zoo has taken native adult silverspot butterflies (*Speryeria zerene hippolyta*) from the wild, induced them to lay eggs, overwintered the larvae in refrigerators and then fed them before releasing them back to the wild to areas where populations of the species already exist. This activity is best described as**

    a. an introduction programme

    b. a reintroduction programme

    c. a population supplementation programme

    d. a translocation programme

4.9i **The coastal management technique whereby land is lost to the sea due to coastal erosion – especially the collapse of cliffs – and human communities are relocated further inland is known as**

    a. managed retreat

    b. land sacrifice

    c. managed withdrawal

    d. planned retreat

4.10i **The coastline of parts of north west England consists of extensive areas of sand exposed to onshore westerly winds for much of the year. Dead Christmas trees are sometimes placed above the high water mark primarily to**

    a. provide habitat for insects

    b. provide food for wading birds

c. encourage sand dune formation

d. provide food for deer

4.11i Coniferous trees are often planted in the watersheds of reservoirs because they

a. increase bird biodiversity

b. intercept rainfall thereby slowing down runoff

c. store large quantities of water in their trunks

d. reduce the albedo

4.12i Which of the following statements about amenity grassland in parks and recreation grounds in many parts of the world is false?

a. They are regularly mown to a short length (close mown)

b. They are high in biodiversity

c. They are treated with chemicals

d. They may contain drainage

4.13i A group of scientists and conservationists led by Josh Donlan have suggested that parts of the Midwest of the United States could be 'restored' by introducing elephants, cheetahs and other species of megafauna similar to those that would now roam these areas if their ancestors had not become extinct (Donlan _et al._, 2006). The introduced species would mimic the roles of the extinct species and they called this process

a. Cretaceous rewilding

b. Permian rewilding

c. Pleistocene rewilding

d. Jurassic rewilding

4.14i Stages in a reintroduction programme for an endangered plant are shown in Table 4.2. Replace the stages labelled K, L and M with the correct information from Table 4.3.

a. A

b. B

c. C

d. D

Table 4.2

Stage	Action
1	Plan and set clear objectives
2	K
3	Propagate plant material
4	Select appropriate reintroduction site(s)
5	L
6	Conduct outplanting
7	M
8	Update protocols based on new information
9	Communicate results to others
10	Maintain habitat
11	Repeat as required

Table 4.3

Stage	A	B	C	D
K	Prepare the site	Assess and interpret results	Obtain source material for reintroduction	Obtain source material for reintroduction
L	Obtain source material for reintroduction	Obtain source material for reintroduction	Assess and interpret results	Prepare the site
M	Assess and interpret results	Prepare the site	Prepare the site	Assess and interpret results

4.15i **Where forest is destroyed to make room for agriculture a zone of forest regrowth may occur at the boundary. Which of the following statements about the tree species and the physical environment at points A and B in Fig.4.1 is false?**

a. Trees at A are likely to be more tolerant of drier conditions than those at B

b. The humidity within the forest is likely to be lower at A than at B

c. The temperature during the day within the forest at A is likely to be lower than that at B

d. Tree species at A are likely to be different from those at B

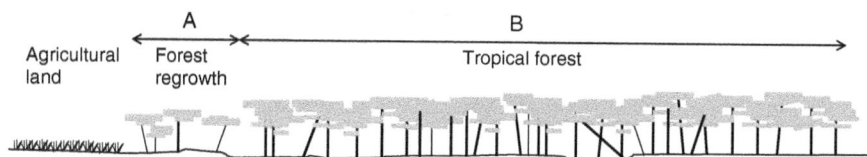

Fig. 4.1.

4.16i In grassland ecosystems, which of the following is not a result of controlled burning?

a. It removes dead plants

b. It prevents the encroachment of trees

c. It releases nutrients to the soil

d. Small mammal species escape damage from fire

4.17i Some plants produce 'recalcitrant seeds'. These are problematic for plant conservationists because they

a. are exceptionally small

b. do not survive drying and freezing

c. are difficult to germinate

d. produce seedlings that are particularly vulnerable to disease

4.18i A real or notional biological community that acts as a model or benchmark for restoration is called a

a. reference ecosystem

b. indicative community

c. aspirational ecosystem

d. target ecosystem

4.19i The cattle breed in Fig. 4.2 is a longhorn and is of conservation interest because

a. it is extinct

b. it is used in conservation grazing

c. it is a rare breed

d. b and c are true

Fig. 4.2.

4.20i **Derelict mine workings may contain piles of the waste materials – consisting of ground rock and process effluents – that were left over when the valuable ore was extracted. This material is collectively known as**

a. overburden

b. tailings

c. drift

d. lode

Advanced

4.1a Goats have been removed from San Clemente Island, California, by using other goats wearing radio-collars to find them so that they could be shot. An individual wearing a radio-collar for this purpose is called a

   a. traitor animal

   b. tracker animal

   c. Judas animal

   d. tracer animal

4.2a The 'single large or several small?' debate relates to the size of

   a. species selected for protection

   b. protected areas

   c. zoo populations used in captive breeding programmes

   d. quadrats used for ecological surveys

4.3a In some nature reserves in Costa Rica horses and cattle are used to mimic the extinct megafauna that would have previously assisted in seed dispersal. Such species are called

   a. ecological homologues

   b. ecological replicates

   c. ecological analogues

   d. ecological mimics

4.4a The tree in Fig. 4.3 is hundreds of years old and typical of those found in ancient forests in Britain. When new forests are created there are no old trees present. The process of artificially ageing trees by injecting them with fungi that cause them to rot in order to create habitat for rare species that depend on decaying ancient trees is known as

   a. vernalisation

   b. veteranisation

   c. maturisation

   d. deteriorisation

Fig. 4.3.

4.5a **A facility that captive breeds rare wildfowl in open pens may remove eggs from the birds' nests and incubate them artificially to**

   a. encourage double-clutching

   b. reduce the loss of eggs and chick to predators

   c. increase the hatching rates of eggs

   d. achieve all of the above

4.6a **Which of the following organisms have been bred by nuns at a convent in Mexico – to produce ingredients for cough syrup – who subsequently used their knowledge to assist with conservation?**

a. Lake Pátzcuaro salamanders (*Ambystoma dumerilii*)

b. Mountain chicken frogs (*Leptodactylus fallax*)

c. Tree frogs (Hylidae)

d. Skinks (Scincidae)

4.7a **Grasses known as metalophytes are grown in some habitats because they**

a. extract unwanted metal from contaminated soil

b. are resistant to heavy metals and will grow on soil that is toxic to other varieties

c. concentrate metals absorbed from the soils and make trace elements available to grazers such as sheep

d. add biodiversity

4.8a **Martin Mere (Fig. 4.4) is a wetland nature reserve in England owned by the Wildfowl and Wetlands Trust (WWT). The water level in the mere (a shallow lake) is**

a. artificially lowered during the breeding season to make a larger shore area available for ground-nesting waterbirds

b. artificially raised to provide shallow water areas around the edge of the mere where ducks and geese can feed

c. raised and lowered at different times of the day to simulate tidal movements

d. maintained at the same depth all year to provide a stable habitat for wading birds

Fig. 4.4.

4.9a The stages in the development of a captive breeding and reintroduction programme are illustrated in Fig. 4.5 (based on Ralls and Ballou (2013)). Match the labels A, B and C in the graph with the terms in Table 4.4.

Table 4.4

Phase	1	2	3	4
A	Growth phase	Founding phase	Capacity phase	Growth phase
B	Capacity phase	Growth phase	Growth phase	Founding phase
C	Founding phase	Capacity phase	Founding phase	Capacity phase

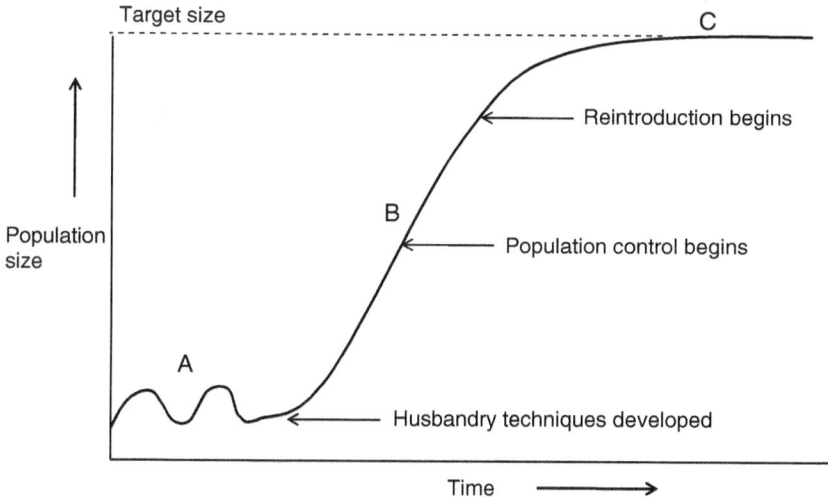

Fig. 4.5.

a. 1

b. 2

c. 3

d. 4

4.10a Grey wolves (*Canis lupus*) were reintroduced to Yellowstone National Park in the United States in 1995 after a long absence during which deer numbers increased and damaged the vegetation. Wolves fed on the deer and in response they moved out of the valleys and gorges where they could easily be predated. The vegetation in these areas re-established, trees grew taller and songbird species and beavers returned. Beaver dams created ponds that provided habitat for otters, muskrats, ducks, reptiles, amphibians and fishes. Wolves killed coyotes so rabbit and mouse populations increased providing more food for hawks, weasels, foxes and badgers. The increase in berries and carrion attracted more bears and their numbers increased. Bears killed deer calves, reinforcing the effect of the wolves on the deer population. This phenomenon is known as

a. an ecological waterfall

b. an ecological chain reaction

  c. a bottom-up trophic cascade

  d. a top-down trophic cascade

4.11a The practice of fostering the natural regeneration of ecosystems by actively removing ecological impediments (e.g. invasive species) and reinstating appropriate biotic and abiotic states (e.g. fire regimes) is known as

  a. ecological maintenance

  b. assisted regeneration

  c. rehabilitation

  d. assisted renewal

4.12a Before a species may be reintroduced into an area it is important for the organisations concerned to consult with local

  a. farmers

  b. residents

  c. politicians

  d. stakeholders

4.13a External exchanges that occur at a level larger than the site under restoration (e.g. energy flows, water, fire, genetic material, animals and seeds) are called

  a. global flows

  b. background flows

  c. landscape flows

  d. scenic flows

4.14a The small size of reintroduced populations may result in an increase in

  a. genetic drift

  b. genetic flow

  c. mutation rate

  d. chromosome deletions

4.15a If a reintroduced population is capable of hybridising with existing wild populations of the same species the result may be

a. outbreeding depression

b. inbreeding depression

c. increased homozygosity

d. genetic drift

4.16a To restore an animal species to a favourable conservation status it may be necessary to remove another (introduced) species with which it hybridises in the wild. For example, in Europe introduced ruddy ducks (*Oxyura jamaicensis*) – a North American species – have been culled to protect native white-headed ducks (*O. leucocephala*) as they produce fertile hybrids. Which of the following is not required for hybridisation to occur?

a. Both species must be closely related

b. The species must have compatible courtship and mating behaviour

c. They must not be mechanically isolated

d. The species must be allopatric

4.17a A phosphate sorbent is most likely to be used to restore a lake that is

a. oligotrophic

b. eutrophic

c. mesotrophic

d. geotropic

4.18a On large developments it is important to store topsoil temporarily before re-spreading on the landscape. The two methods used are referred to as

a. dry and wet stockpiling

b. compacted and non-compacted stockpiling

c. aerobic and anaerobic stockpiling

d. vertical and horizontal stockpiling

4.19a **Which of the following is least likely to be a result of the long-term stockpiling of soil?**

    a. A reduction in aggregate stability (the ability to resist degradation by water, wind, etc.)

    b. An increase in potential for mycorrhizal infection

    c. A change in pore size

    d. An increase in bulk density (weight of soil in a given volume)

4.20a **The term 'field propagation and release' refers to a type of**

    a. recovery programme for rare seed-bearing plants

    b. captive breeding programme for animals in their natural habitat

    c. restoration programme for old fields

    d. forest restoration project

5 Agriculture, Forestry and Fisheries Management

This chapter consists of questions about ecological aspects of agriculture, forestry and fisheries management, including game ranching and whaling.

Foundation

5.1f Marine animals (e.g. cetaceans) unintentionally caught in fishing nets during commercial fishing operations are referred to as

    a. accidental catch

    b. bycatch

    c. incidental catch

    d. residual catch

5.2f The cultivation of marine organisms for food and other products in seawater is most accurately described as

    a. aquaculture

    b. aquafarming

    c. fish farming

    d. mariculture

5.3f A crop grown for sale rather than to feed local people is called a

    a. subsistence crop

    b. food crop

c. cash crop

d. currency crop

5.4f   Hydroponics is a system of growing plants without

a. insecticides

b. water

c. soil

d. fertilisers

5.5f   Coppicing and pollarding are practices employed in

a. silviculture

b. agriculture

c. horticulture

d. aquaculture

5.6f   Wind can cause water stress in crop plants because it

a. increases transpiration

b. damages stems

c. reduces water percolation

d. encourages denitrification

5.7f   Crushed limestone is applied to some agricultural soils to increase the levels of

a. phosphorus

b. nitrogen

c. sulphur

d. calcium

5.8f   Many power stations in Europe, North America, Japan and Russia have associated fish farms. This is because the power stations can supply the farms with

a. inexpensive electricity

b. heated water

c. nutrient-rich water

d. oxygenated water

5.9f **Transhumance is a traditional practice that involves**

    a. the herding of reindeer populations by nomadic people

    b. keeping livestock in bomas at night as protection from predators

    c. managing flocks of sheep on common land

    d. the seasonal movement of livestock between fixed summer and winter pastures

5.10f **A fish ladder is most likely to be found associated with**

    a. a bridge across a river

    b. an obstruction across a river, e.g. a dam or weir

    c. a fish farm

    d. an aquarium

5.11f **The rearing of several fish species with different feeding habits together in ponds to maximise the utilisation of natural food is known as**

    a. polyculture

    b. multiculture

    c. aquaculture

    d. mariculture

5.12f **The Rainforest Trust protects tropical forests primarily by**

    a. fencing large areas of forest

    b. relocating indigenous forest peoples

    c. draining marshland and planting trees

    d. purchasing forest land

5.13f **Which of the following statements about trees is false?**

    a. Hard woods are broad-leaved trees

    b. Timber from hardwoods has a higher density than that from softwoods

    c. Softwoods grow slower than hardwoods

    d. Timber from hardwoods is stronger and more durable than that from softwoods

5.14f **Experiments were conducted on game ranching in Africa in the 1960s by the American ecologist**

a. Raymond Dasmann

b. Eugene Odum

c. Robert MacArthur

d. Daniel Janzen

5.15f **The label in Fig. 5.1 can be seen on some bananas and has had some of its text redacted so that the organisation responsible is hidden. It was produced by an international NGO that promotes environmental certification for sustainable forestry, agriculture and tourism. The hidden text says**

a. Tropical Alliance

b. Rainforest Union

c. Rainforest Alliance

d. Forest Coalition

Fig. 5.1.

5.16f **Many crop plants require a period of exposure to low temperatures to switch from vegetative to reproductive growth. This process is called**

a. ventilation

b. vernalisation

c. valorisation

d. volatilisation

5.17f **The incorporation of trees and shrubs into agricultural land used for crops or livestock is known as**

a. silviculture

b. agroforestry

c. agroecology

d. agronomy

5.18f **Ash dieback is a serious disease of ash trees (*Fraxinus* spp.) caused by**

a. a fungus

b. a bacterium

c. a virus

d. an insect parasite

5.19f **Which of the following types of tree is a hardwood?**

a. Redwood

b. Beech

c. Spruce

d. Pine

5.20f **The process that results in the genetic adaptation of animals controlled by humans, for example for food and transportation, is called**

a. cross-breeding

b. inbreeding

c. domination

d. domestication

Intermediate

5.1i   Which of the following marine areas does not have a whale sanctuary?

   a. Antarctica

   b. Indian Ocean

   c. Southern Ocean

   d. North Sea

5.2i   Game ranching has been developed in several African countries. Some fencing systems allow game animals to enter different grazing areas – often attracted to them by the provision of water – and move between them by opening and closing gates. Such systems facilitate

   a. transitional grazing

   b. circular grazing

   c. rotational grazing

   d. sequential grazing

5.3i   Intercropping is the cultivation of a crop among plants of a different type. This may

   a. increase the yield on a given piece of land

   b. reduce the dispersal of pest organisms

   c. increase the biodiversity of insects and soil organisms

   d. facilitate all of the above, depending on the specific crops involved

5.4i   A fish net that has a very fine mesh and is capable of catching very small, young fish that have not had the opportunity to reproduce is called a

   a. seine net

   b. mist net

   c. cannon net

   d. fyke net

5.5i The graph below (Fig. 5.2) shows the relationship between fishing effort and the catch landed (yield) from a marine fishery. Which of the labelled points on the graph indicates the maximum sustainable yield?

   a. A

   b. B

   c. C

   d. D

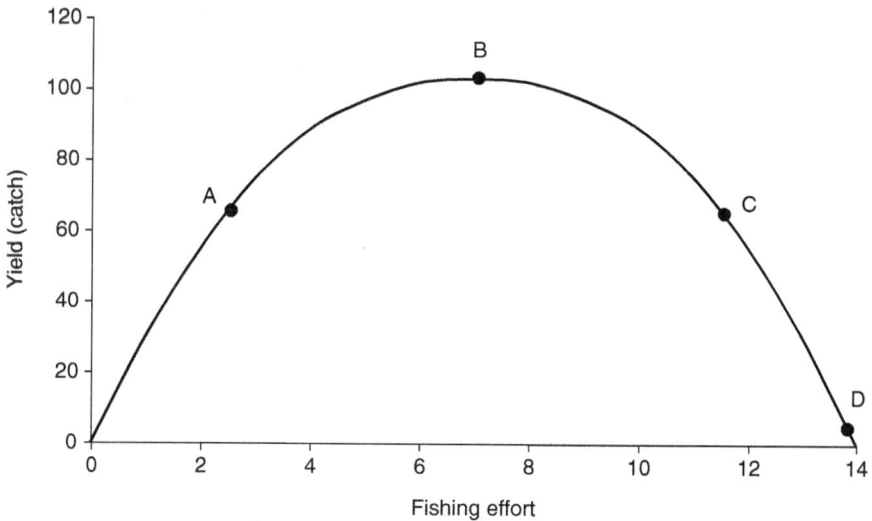

Fig. 5.2.

5.6i The D-value of a grass is a measure of its

   a. dry matter yield

   b. protein content

   c. growth rate

   d. digestibility

5.7i The oxygen content of the soil is reduced by

   a. fertiliser application

   b. waterlogging

   c. ploughing

   d. subsoiling

5.8i A 'blue whale unit' was

    a. a means by which whaling quotes were defined in the past

    b. a whale research centre in Antarctica

    c. a fixed quantity of blue whale meat

    d. the name of a type of whaling ship that specialised in processing blue whales

5.9i The list of rare farm animal breeds published annually by the Rare Breeds Survival Trust is called the

    a. Red List

    b. Watchlist

    c. Herd List

    d. Yellow List

5.10i Chlorosis is a condition in plants that results in a lack of chlorophyll caused by

    a. a deficiency of magnesium

    b. poor soil drainage

    c. exposure to sulphur dioxide

    d. any of the above

5.11i Which element is added to the soil by the inclusion of legumes in a crop rotation?

    a. Phosphorus

    b. Sulphur

    c. Nitrogen

    d. Potassium

5.12i One of the reasons for the widespread decline in honey-bees is known as

    a. colony collapse disorder

    b. colony collapse disease

    c. honeybee decline disorder

    d. honeybee decline phenomenon

5.13i Which of the following is not a rare breed of pig?

    a. Middle white

    b. Gloucestershire old spot

    c. Golden Guernsey

    d. British saddleback

5.14i Which of the following results in an increase in nitrogen in the soil nitrogen pool?

    a. Rainfall

    b. Denitrification

    c. Immobilisation

    d. Leaching

5.15i Which of the following changes to fish catches would you expect to occur after mesh regulation which increased the size of the holes in fishing nets?

    a. Fewer very young fish caught

    b. More older fish caught

    c. No very young fish and a higher proportion of larger fish caught

    d. A greater weight of young fish caught

5.16i The method of managing fields whereby cattle are allowed to feed in one section while being excluded from the remainder of the area by an electric fence is known as

    a. sectional grazing

    b. strip grazing

    c. exclusion grazing

    d. zonal grazing

5.17i Waterlogging of agricultural soil may result in

    a. poor seedling development

    b. accelerated denitrification of nitrates

    c. the build-up of plant-toxic gases such as methane

    d. all of the above

5.18i Game ranching is more productive than cattle farming – in terms of meat production per unit area – because

    a. game animals grow faster than cattle

    b. different game species eat a wide range of plants while cattle are selective grazers

    c. game animals produce more meat than cattle

    d. more cattle than game animals are lost to parasitic infections

5.19i In aquaculture the term 'growing on' refers to growing fish

    a. on pond weed

    b. on to marketable size

    c. on manufactured fish food

    d. on land in tanks

5.20i There are over 500 species of sea lice. They are commercially important ectoparasites of marine fish species and are

    a. copepods

    b. amphipods

    c. ostracods

    d. hexapods

Advanced

5.1a Modern intensive farming methods include spraying fields with chemical fertilisers, herbicides and insecticides that may be blown onto adjacent land by the wind; a phenomenon known as spray drift (Fig.5.3). Which of the following adjustments to spraying equipment is likely to reduce spray drift?

    a. A reduction in spray pressure

    b. The use of a nozzle that produces a finer spray

    c. An increase in sprayer ground speed

    d. Raising the height of the spray boom

Fig. 5.3.

5.2a. Which of the following lists of terms used in forestry is correctly paired with their definitions in Table 5.1?

Table 5.1

Term	Definition			
	A	**B**	**C**	**D**
Selective cutting	Harvesting of a portion of a forest	Trees individually harvested from a diverse forest	Trees individually harvested from a diverse forest	Felling of an entire forest
Clear cutting	Felling of an entire forest	Felling of an entire forest	Harvesting of a portion of a forest	Trees individually harvested from a diverse forest
Strip or stand cutting	Trees individually harvested from a diverse forest	Harvesting of a portion of a forest	Felling of an entire forest	Harvesting of a portion of a forest

a. A

b. B

c. C

d. D

5.3a   **Which of the following statements about wildfires is false?**

a. Fires burn more rapidly when temperatures are high and humidity is low

b. New spot fires can be ignited by embers blown for miles by the wind

c. Fires burn more rapidly when moving down a slope because wind moves more rapidly down slopes

d. Wind increases the oxygen supply and may cause a fire to grow rapidly

5.4a   **Territory size in male springbok (*Antidorcas marsupialis*) is approximately 0.22km². The number of rams that can be kept on a game ranch of 5km² is approximately**

a. 15

b. 19

c. 23

d. 27

5.5a   **Which of the following fisheries has collapsed since the 1990s?**

a. Atlantic northwest cod fishery

b. Newfoundland cod fishery

c. British Columbia salmon fishery

d. All of the above.

5.6a   **Genetically-modified (GM) crops possess a genome into which genes have been inserted from another species. By which of the following means has this been achieved?**

i.   Artificially using bacteria

ii. Naturally from bacteria in the environment

iii. Artificially using viruses

iv. Using small metal particles

a. i, ii and iii

b. i, iii and iv

c. ii, iii and iv

d. i, ii, iii and iv

5.7a **Which of the following statements about the shelterwood method of harvesting timber is false?**

a. It leaves a large percentage of mature trees standing to allow the forest to regenerate from tree seedlings

b. It makes the forest more vulnerable to storm damage

c. It allows the forester to determine which tree species thrive by controlling the amount of shade by adjusting the proportion of mature trees felled

d. It is usually achieved by a single round of tree cutting

5.8a **The table below (Table 5.2) illustrates the effect of plant density on yield in wheat.**

Table 5.2

Plant density (plants/m²)	Yield/plant (g)	Yield (g/m²)
7	24.7	173
154	1.5	231
447	0.4	179

Which of the following statements is false?

a. At the highest density interspecific competition for light and nutrients reduces yield

b. The yield/plant decreases with plant density

c. Tripling the plant density that produces a yield of 1.5 g/plant will cause a reduction in overall yield (g/m²)

d. The highest yield (g/m²) is achieved at a plant density of 154 plants/m²

5.9a In crop production, the ratio between the investment in seed and the seed yield is a measure of productivity and is known as the

a. crop ratio

b. seed ratio

c. yield ratio

d. productivity ratio

5.10a Methyltestosterone is added to the diet of some cultured fish to

a. prevent disease

b. convert females to males

c. treat fungal diseases

d. improve food assimilation

5.11a The Park Grass Experiment is a study established to examine the effects of manures and fertilisers on hay yields and various aspects of farmland ecology at the

a. Boorowa Agricultural Research Station, Australia

b. Federal Agricultural Research Centre, Germany

c. Rothamstead Experimental Station, England

d. Connecticut Agricultural Experiment Station, United States

5.12a Sustained overfishing can lead to a situation whereby a decrease in the breeding population causes a reduced production and survival of eggs or offspring resulting in an inability of the population to sustain itself known as a

a. critical dispensation

b. critical biomass

c. critical depression

d. critical depensation

5.13a **A major American restaurant chain announced in 2020 that it was adding lemongrass (*Cymbopogon*) to the diet of some of the cows used to produce its beefburgers to**

    a. improve animal welfare

    b. improve the flavour of the meat

    c. reduce the fat content of the meat

    d. reduce methane emission from the cows

5.14a **Leaf area duration is a measure of crop 'leafiness' and is measured in terms of**

    a. area

    b. time

    c. volume

    d. biomass

5.15a **Table 5.3 shows the number of mites/5cm^2 found on a crop grown in glasshouses (greenhouses) under two different control methods. These pests may be controlled by pesticides (chemical control) or by using predatory mites (biological control). Which is the most likely explanation for the data in the table below?**

    a. Biological control was used in both glasshouses

    b. Chemical control was used in both glasshouses

    c. Biological control was used in glasshouse 1 and chemical control in glasshouse 2

    d. Chemical control was used in glasshouse 1 and biological control in glasshouse 2

Table 5.3

Time (days)	Number of pest mites/5cm²	
	Glasshouse 1	Glasshouse 2
15	2.1	2.9
30	14.3	2.6
45	1.9	2.7
60	20.3	2.8
75	1.2	2.4
90	15.1	2.2
105	1.7	2.5
120	24.5	1.9
135	1.2	2.3
150	4.0	2.1
165	0.9	2.4
190	9.7	2.6
205	1.1	2.2

5.16a Ammonia is the main excretory product released by aquatic animals and its accumulation in water will cause undesirable changes in pH. In aquaculture, the natural conversion of ammonia to nitrite by bacteria may be accelerated by

a. increasing the temperature

b. reducing the temperature

c. adding carbon dioxide

d. oxygenation

5.17a The term 'pharming' refers to which of the following?

a. The use of genetic engineering to insert one or more genes that code for a pharmaceutical into a host plant or animal thereby creating a genetically-modified organism

b. The use of pharmaceuticals to improve the health of livestock

c. The use of pharmaceuticals to improve crop growth and plant health

d. The use of pharmaceuticals in all aspects of plant and animal husbandry in agriculture

5.18a **Which of the following actions would make a forest more resilient to fire?**

i. Removing dry branches from the forest floor

ii. Pruning branches from the tree base

iii. Removing small trees

iv. Removing large logs from the forest floor

v. Removing leaves from the forest floor

vi. Increasing the space between trees

a. i, ii and vi

b. i, ii, iii and vi

c. i, ii, iii and iv

d. ii, iii, iv and vi

5.19a **In aquaculture, hypophysation is the injection of pituitary hormones into fish whose gonads have already matured in order to**

a. induce spawning

b. improve food assimilation

c. reduce intraspecific aggression

d. stimulate growth

5.20a **In agriculture drones (unmanned aerial vehicles) have been used to**

a. deliver pollen to crops

b. monitor crop growth

c. detect pest and fungal outbreaks in crops

d. accomplish all of the above

6 Pest, Weed and Disease Management

This chapter contains questions about pests and weeds and their control, invasive species, parasites, and animal and plant diseases.

Foundation

6.1f Chytridiomycosis is a fungal disease that threatens the survival of populations of

    a. birds

    b. amphibians

    c. reptiles

    d. bees

6.2f Which of the following is most likely to be effective in trapping cockroaches?

    a. Sticky traps

    b. Longworth traps

    c. Sherman traps

    d. Light traps

6.3f A drug used to treat animals infected with tapeworms is called an

    a. antihistamine

    b. anthelmintic

c. antibiotic

d. anticoagulant

6.4f   The chemical 2,4-D is

a. an insecticide

b. a fertiliser

c. a fungicide

d. a herbicide

6.5f   A disease that may be transmitted between animals and humans is called a

a. zoonosis

b. pandemic

c. contagion

d. zoophobia

6.6f   A chemical that inhibits the feeding behaviour of an insect pest is called

a. a pheromone

b. a repellant

c. an antioxidant

d. an antifeedant

6.7f   Chronic wasting disease (CWD) is an emerging infectious disease that affects members of the mammalian family

a. Cervidae

b. Bovidae

c. Suidae

d. Felidae

6.8f   Ebola is a viral disease that is thought to have been transmitted to humans via

a. primates used in the bushmeat trade

b. insect bites in tropical forests in Central Africa

c. bats in South America

d. contact with domestic animals in West Africa

6.9f The control of pests in urban areas by combining the use of improved sanitation, habitat modification, trapping, pesticides and biological control is known as

a. consolidated pest management

b. amalgamated pest control

c. integrated pest management

d. holistic pest control

6.10f Dutch elm disease is caused by

a. a fungus spread by beetles

b. a virus spread by flies

c. a bacterium spread by aphids

d. a fungus spread by birds

6.11f The Great Irish Famine of 1845-49 was caused by failure of the potato crop resulting from infection by

a. *Plasmopara viticola*

b. *Fusarium oxysporum*

c. *Phytophthora infestans*

d. *Venturia inaequalis*

6.12f Which animals are responsible for the vast majority of human deaths from rabies?

a. Cats

b. Dogs

c. Foxes

d. Wolves

6.13f Elephantiasis is a disease of humans that causes swelling of the legs as a result of

a. blockage of lymphatic vessels by a leech

b. blockage of lymphatic vessels by a nematode

   c. blockage of blood vessels by an insect larva

   d. blockage of blood vessels by a cestode

6.14f Regular farm operations designed to destroy pests or prevent them from doing economic damage that do not require special equipment or skills are collectively referred to as

   a. cultural control

   b. natural control

   c. indirect control

   d. ecological control

6.15f A market where live wild animals are sold as food, especially in Southeast Asia, are a source of diseases that may be transmitted to humans. Such markets are often referred to as

   a. animal marts

   b. animal bazaars

   c. dry markets

   d. wet markets

6.16f If a foot-and-mouth disease outbreak occurs in a zoo, which of the animals in the following list would be susceptible?

   i. Elephants

   ii. Buffalo

   iii. Alpacas

   iv. Giraffes

   v. Deer

   vi. Antelopes

   vii. Lions

   a. i, iii, iv, v and vi

   b. i, ii, iii, iv, v and vi

   c. ii, iv, v and vi

   d. ii, iii, iv, v and vii

6.17f The organism known as TMV causes disease in which of the following commercially grown plants?

    a. Tulips

    b. Tobacco

    c. Thyme

    d. Turnips

6.18f Complete the following sentence using one of the terms listed below: 'Fields of agricultural crops (monocultures) are susceptible to disease because the individual plants are the same species and all have similar ....'

    a. appearances

    b. genotypes

    c. nutrient requirements

    d. physiology

6.19f The disease myxomatosis was introduced into Australia to control populations of

    a. dingoes (*Canis lupus dingo*)

    b. red foxes (*Vulpes vulpes*)

    c. rabbits (*Oryctolagus cuniculus*)

    d. feral cats (*Felis catus*)

6.20f The structures in Fig. 6.1 are caused by

    a. a fungus

    b. a parasitic nematode

    c. a bacterial infection

    d. an insect

Fig. 6.1.

Intermediate

6.1i Sleeping sickness is

a. caused by *Anopheles* and transmitted by *Plasmodium*

b. caused by *Plasmodium* and transmitted by *Anopheles*

c. caused by *Trypanosoma* and transmitted by *Glossina*

d. caused by *Glossina* and transmitted by *Trypanosoma*

6.2i Queleas (Ploceidae) are of economic importance in Africa because they are

a. parasites that cause disease in cattle

b. grain-eating bird pests

c. insect pests of coffee plants

d. beetles that act as vectors for a fungus that damages trees

6.3i A chemosterilant protects crops by

a. causing temporary or permanent sterility in pest insects

b. killing rodents that feed on and damage them

c. inhibiting the growth of harmful bacteria in the soil

d. sterilising leaves to prevent fungal growth

6.4i What is the leading cause of mosquito-borne disease in the continental United States?

    a. Japanese encephalitis virus

    b. West Nile virus

    c. Dengue virus

    d. Zika virus

6.5i Which of the following is/are not a second generation insecticide?

    a. Carbamates

    b. Hydrogen cyanide

    c. Chlorinated hydrocarbons

    d. Organophosphates

6.6i Which of the following is a synthetic auxin?

    a. 2,4-D

    b. Glyphosate

    c. Paraquat

    d. Mesotrione

6.7i Planting a second crop near the production crop that is more attractive to pests than the production crop is called

    a. relay cropping

    b. push-pull cropping

    c. repellent cropping

    d. trap cropping

6.8i Since the 1990s populations of the Tasmanian devil (*Sarcophilus harrisii*) have been depleted by

    a. rabies

    b. a contagious cancer

    c. anthrax

    d. tick fever

6.9i Crop plants that are growing at the wrong time or in the wrong place, for example seedlings arising from grain shed during combining growing in a field of potatoes, are called

    a. volunteers

    b. conscripts

    c. interns

    d. recruits

6.10i Which of the following is not an invasive species in Australia?

    a. Cane toad (*Rhinella marina*)

    b. Donkey (*Equus asinus*)

    c. Dromedary (*Camelus dromedarius*)

    d. Badger (*Meles meles*)

6.11i When red scale (*Aonidiella aurantii*) is controlled in citrus fruits using insecticides the natural enemies of the white wax scale (*Gascardia destructor*) are killed, causing its numbers to increase so that it too becomes a pest. This is known as a

    a. facultative pest outbreak

    b. secondary pest outbreak

    c. subordinate pest outbreak

    d. incidental pest outbreak

6.12i Bt corn contains a gene from the bacterium *Bacillus thuringiensis* (hence Bt corn) that produces a protein that kills the larvae of Lepidoptera. It is an example of a

    a. chimera

    b. clone

    c. pangenic plant

    d. transgenic plant

6.13i The caterpillar moth (*Cactoblastis cactorum*) has been successfully used to control

    a. Himalayan balsam (*Impatiens glandulifera*)

    b. Japanese knotweed (*Reynoutria japonica*)

    c. prickly pear (*Opuntia* spp.)

    d. giant hogweed (*Heracleum mantegazzianum*)

6.14i **In some tropical countries legislation has been introduced to prevent the growing of certain crops at particular times of the year to control pests. For example, in Kenya, after a cotton crop has been harvested, plants are uprooted and burned to prevent a build-up of the pink bollworm (*Pectinophora gossypiella*) and seeds cannot be planted until the following rains. The period when the plant may not legally be grown is known as a**

    a. disinfection season

    b. pest control season

    c. close season

    d. open season

6.15i **Some insect pests can be controlled by SIT. This acronym stands for**

    a. sterile insecticide technology

    b. sterile insect technique

    c. sprayed insecticide technology

    d. superior insecticide technology

6.16i **The life cycle of the tapeworm *Echinococcus vogeli* is illustrated in Fig. 6.2. The definitive host is**

    a. the bush dog (*Speothus venaticus*)

    b. the paca (*Cuniculus* spp.)

    c. the domestic dog (*Canis familiaris*)

    d. the bush dog (*Speothus venaticus*) and the domestic dog (*Canis familiaris*)

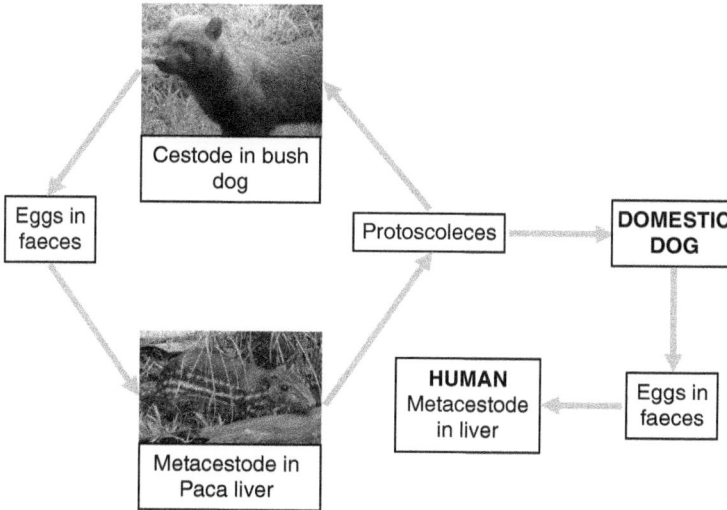

Fig. 6.2.

6.17i **Which of the following is not an important disease of cattle?**

    a. Rabies

    b. Anthrax

    c. Tuberculosis

    d. Newcastle disease

6.18i **Which animals have been particularly implicated as the source of severe acute respiratory virus (SARS) in China?**

    a. Bats

    b. Birds

    c. Rodents

    d. Pigs

6.19i **Rinderpest kills livestock and wildlife such as giraffes, buffalo, wildebeest, warthogs and antelopes. It is caused by a**

    a. bacterium

    b. protozoan

    c. virus

    d. prion

6.20i Which of the following is a major invasive plant in Lake Victoria?

   a. Water hyacinth (*Eichhornia crassipes*)

   b. Water lettuce (*Pistia stratiotes*)

   c. Water clover (*Marsilea* spp.)

   d. Turtleweed (*Batis maritima*)

Advanced

6.1a Chimpanzees (*Pan* sp.) and gorillas (*Gorilla* sp.) infected with simian immunodeficiency virus (SIV) that entered the bushmeat trade were probably responsible for a human disease known as

   a. SARS

   b. MERS

   c. EID

   d. HIV

6.2a Warfarin-resistant rodents may also be resistant to other anti-coagulant rodenticides such as chloro-phacinone. This phenomenon is called

   a. conferred-resistance

   b. ultra-resistance

   c. cross-resistance

   d. multi-resistance

6.3a What is the LD_{50} of the insecticide whose toxicity is shown in Fig. 6.3?

   a. 58%

   b. 47 mg/m^2

   c. 80 mg/m^2

   d. 71%

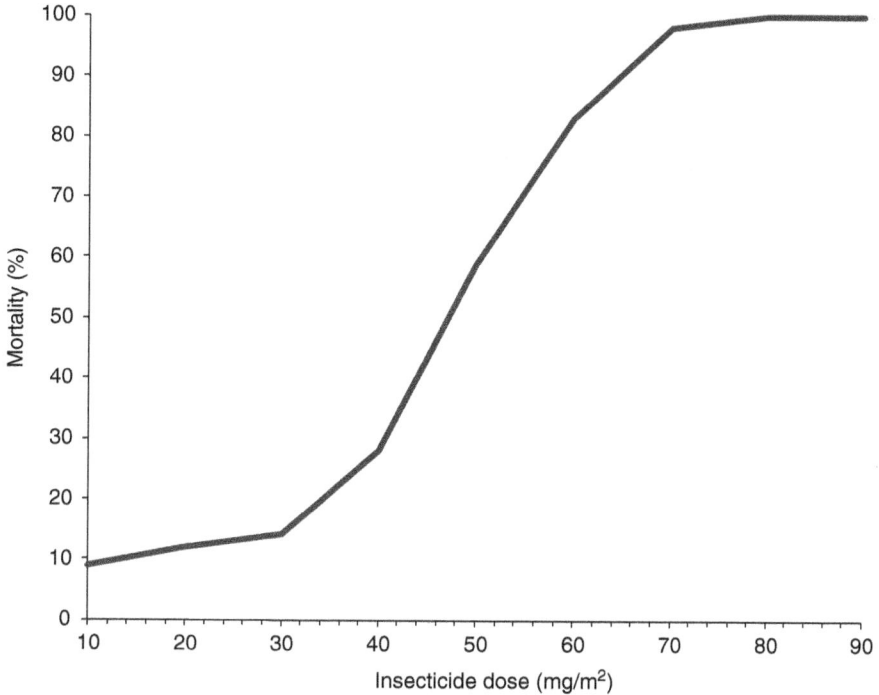

Fig. 6.3.

6.4a **The use of insecticides capable of killing a broad spectrum of insect species may lead to an outbreak of an insect pest as a result of the chemical killing a relatively greater number of natural enemies than pests. The pest may increase and reinvade a sprayed area and rapidly increase in numbers in a phenomenon known as**

   a. recovery

   b. revival

   c. resurgence

   d. re-emergence

6.5a **Occasionally a natural enemy may make an appearance in the environment, unaided by humans, and become an effective mortality factor that controls a pest. This phenomenon is known as**

   a. fortuitous biological control

   b. classical biological control

c. accidental biological control

d. unintentional biological control

6.6a **In the food chain below, which of the following is true?**

        A   B   C

Erinnyis ello → *Cryptophion* sp. → *Spilochalcis* sp.

a. A is a hyperparasitoid; B is a parasitoid; C is the host

b. A is the host; B is a vector; C is a hyperparasitoid

c. A is the host; B is a hyperparasitoid; C is an endoparasite

d. A is the host; B is a parasitoid; C is a hyperparasitoid

6.7a **Some crops are capable of synthesising weed-inhibiting chemicals that are produced by the actively growing plant or arise from residues after its death. This phenomenon is known as**

a. phytopathy

b. chemopathy

c. genopathy

d. allelopathy

6.8a **Collision with birds, especially during take-off and landing, can cause serious damage to aircraft. Which of the following types of birds are responsible for the majority of bird strikes?**

a. Swans and eagles

b. Ducks and gulls

c. Owls and swans

d. Swans and ducks

6.9a **Which of the following have been used in the biological control of weeds?**

i. Mites

ii. Nematodes

iii. Parasitic plants

iv. Fungi

v. Insects

a. i, ii and v

b. ii, iv and v

c. ii, iii, iv and v

d. i, ii, iii, iv and v

6.10a Firewood entering the United States from Canada must be heat treated and certified free of

a. house longhorn beetle (*Hylotropes bajulus*)

b. emerald ash borer (*Agrilus planipennis*)

c. death watch beetle (*Xestobium rufovillosum*)

d. furniture beetle (*Anobium punctatum*)

6.11a Which of the following lists (Table 6.1) contains examples of organochlorine and organophosphate insecticides?

Table 6.1

A	B	C	D
DDT	Heptachlor	Dieldrin	Lindane
Malathion	DDT	Aldrin	Heptachlor
Lindane	Dieldrin	Heptachlor	Dieldrin
Aldrin	Lindane	DDT	Aldrin

a. A

b. B

c. C

d. D

6.12a The likelihood that an insect pest will become resistant to an insecticide may be reduced by

a. using several different classes of pesticides in a long-term rotation

b. mixing two insecticides together

c. increasing the concentration of the insecticide with successive applications

d. increasing the application frequency of the insecticide

6.13a Which of the following statements about fascioliasis is false?

    a. It is caused by the cestode *Fasciola hepatica*

    b. Sheep are one of the main primary hosts of the parasite that causes the disease

    c. Adult parasites infect the heart of the primary host

    d. Keeping livestock away from wet fields may help to control the disease

6.14a Citrus oils are the basis of some

    a. organic insecticides

    b. organic acaricides

    c. organic fertilisers

    d. organic insecticides and organic acaricides

6.15a A phytosanitary certificate is an inspection certificate used to show that

    a. a particular shipment of plants is free from harmful pests and plant diseases

    b. a particular shipment of plants or plant products is free from harmful pests and plant diseases

    c. a particular shipment of plants or plant products has been treated for plant diseases

    d. a particular shipment of plants or plant products travelling between countries is free from harmful pests and plant diseases

6.16a In the European Union, which agency is responsible for the risk assessment of plant protection products?

    a. The European Food Safety Authority

    b. The European Centre for Disease Prevention and Control

    c. The European Environment Agency

    d. The European Medicines Agency

6.17a **In the United States, the agency responsible for keeping its agricultural industries free from pests and diseases is known by the acronym**

  a. APHID

  b. APHIS

  c. ADAS

  d. USDA

6.18a **Animal and plant diseases and pests whose presence must, by law, be reported to the appropriate government authorities are often described as**

  a. notifiable

  b. reportable

  c. advisable

  d. disclosable

6.19a **Controlling weeds by weed pulling, mowing, mulching or tillage is known as**

  a. biological weed control

  b. destructive weed control

  c. mechanical weed control

  d. mechanised weed control

6.20a **Soil solarisation is a method of controlling**

  a. plant parasites

  b. livestock parasites

  c. insect pests

  d. weeds

7 Urban Ecology and Waste Management

This chapter contains questions about the structure and functioning of urban ecosystems and the management of the waste they produce.

Foundation

7.1f The name given to areas of woodland and agricultural land on the edge of an urban area where building is restricted is often called

    a. green belt

    b. green band

    c. green ring

    d. green girdle

7.2f A brownfield site may be

    a. an area of land occupied by a derelict cotton mill

    b. the site of a disused coalmine

    c. a neighbourhood where Victorian housing has been demolished

    d. any of the above

7.3f Graveyards in the United Kingdom often contain communities of plants that have disappeared from the surrounding area. These are known as

    a. vestigial communities

    b. residual communities

c. remnant communities

d. relic communities

7.4f **With regard to the ecological requirements of birds, city centres have been likened to**

a. forests

b. deserts

c. cliffs

d. moorland

7.5f **Dr Jones lives in a rural area in Bolton, on the edge of Greater Manchester in the United Kingdom. It takes her approximately 30 minutes to drive the 15 miles to work in the city, which is at about the same height above sea level as her home. When she drives to work on warm days in summer Dr Jones notices that the outside temperature recorded in her car increases as she approaches the built-up area of the city. This phenomenon is known as the**

a. urban heat effect

b. heat island effect

c. greenhouse effect

d. urban warming effect

7.6f **A wild species of animal or plant that lives in association with humans and the habitats they have created is called**

a. a synanthrope

b. an anthophile

c. a biophile

d. an anthropomorph

7.7f Large flocks of starlings (*Sturnus vulgaris*) live in some areas, roosting in trees, on the window ledges of tall buildings and on the metal superstructure of piers at seaside resorts. Prior to roosting they perform impressive coordinated aerial displays in close formation known as

a. murmurations

b. numerations

c. machinations

d. susurrations

7.8f The following is a list of methods by which plant seeds may be transported within an urban ecosystem:

i. By wind

ii. On vehicle tyres

iii. By water

iv. In mud on the feet of birds

v. In mud on footwear

vi. In the fur of mammals

vii. In the faeces of animals

viii. In topsoil and rubble

Which of the methods listed are anthropogenic?

a. i, iii, iv, vi, vii and viii

b. ii, v, viii

c. ii, v, vii and viii

d. iv, vi and vii

7.9f Which of the following is not a characteristic of a green roof (Fig. 7.1)?

a. It improves air quality

b. It provides sound insulation

c. It increases the rate of stormwater runoff

d. It regulates the interior temperature of the building

Fig. 7.1.

7.10f **Mosquitoes transmit a number of diseases including malaria, yellow fever, dengue and filariasis. In urban areas they are likely to breed in**

    a. compost heaps

    b. garden ponds

    c. refuse dumps

    d. lawns

7.11f **The albedo of an area of the Earth's surface is a measure of the solar energy reflected from a surface. A white surface reflects most of the energy it receives so it has a high albedo. Tree clearance causes**

    a. an increase in albedo

    b. a decrease in albedo

    c. no change in albedo

    d. a doubling of the albedo

7.12f **Trees and other plants have colonised the abandoned building in Fig. 7.2 because of their powers of**

    a. diffusion

    b. dispersion

    c. dispersal

    d. distribution

Fig. 7.2.

7.13f **Complete the following sentence using one of the words listed below: 'As a result of the use of salt to de-ice roads in Britain .... plant species have invaded roadside verges.'**

    a. agricultural

    b. maritime

    c. horticultural

    d. cultivated

7.14f Private gardens are an important habitat for wildlife. Which of the following European countries has the highest proportion of houses with gardens?

a. United Kingdom

b. The Netherlands

c. France

d. Spain

7.15f A bioswale is

a. a small vegetated area in an urban location that is managed to attract birds

b. a type of grass roof that supports a very high diversity of plants

c. a type of roadside verge found in urban areas that has been seeded with wild flowering plants

d. a narrow vegetated drainage trough that intercepts and impedes the flow of storm water and acts as a biofilter

7.16f The 'Aberfan Disaster' that occurred in South Wales in 1966 was the result of

a. pollution of a local river

b. derelict land contamination with heavy metals

c. the collapse of a colliery spoil tip

d. the structural failure of a reservoir dam

7.17f 'Garden cities', new town developments which promoted 'public improvement' and were self-contained communities that included housing, industry, orchards, offices and shops, were first developed in England in the

a. 1890s

b. 1900s

c. 1920s

d. 1930s

7.18f In a sewerage system, infiltration is

a. the flow received into the sewer from domestic premises

b. the flow received into the sewer from industrial premises

c. the water dumped into the sewer from improper connections

d. the groundwater that enters the sewer through leaks in the pipes

7.19f   Bird spikes are most likely to be found on

a. trees

b. buildings

c. hedges

d. vehicles

7.20f   In urban environments there are many opportunities for people to come into close contact with animals that may be carrying zoonotic diseases. Which of the following is the most important reservoir of *Yersinia pestis*, the bacterium that causes plague?

a. Pigeons

b. Dogs

c. Cats

d. Rodents

Intermediate

7.1i   Blair (1996) has defined 'urban avoiders' as

a. individual animals that avoid urban areas

b. species that reach their highest densities at the most natural sites

c. plants that do not thrive in cities

d. people who prefer to live in rural environments

7.2i   The home range of foxes (*Vulpes vulpes*) living in urban areas

a. tend to be smaller than those in rural areas

b. tend to be larger than those in rural areas

c. are approximately the same as those in rural areas

d. have not been scientifically studied

7.3i **The activated sludge process is used to treat wastewater (sewage) and involves the mixing and aeration of tanks of sewage and microbes. This process was invented in**

a. Frankfurt, Germany

b. Amsterdam, The Netherlands

c. Chicago, United States

d. Manchester, United Kingdom

7.4i *Dittrichia viscosa* **is a flowering plant found along roadsides and railway tracks in Mediterranean countries. Plant species that live in disturbed habitats are referred to as**

a. littoral species

b. benthic species

c. lotic species

d. ruderal species

7.5i **Which of the following species has not adapted to living and breeding in cities?**

a. Peregrine falcon (*Falco peregrinus*)

b. Common buzzard (*Buteo buteo*)

c. European starling (*Sturnus vulgaris*)

d. European stork (*Ciconia ciconia*)

7.6i **The simplest way to maximise the biodiversity in an English garden is to**

a. construct a pond

b. use inorganic fertiliser to increase plant growth

c. introduce more three-dimensional complexity in the vegetation

d. use insecticides to kill predatory insects

7.7i **Waste from human settlements that cannot be recycled is often disposed of in landfill sites. A landfill site should be lined with a layer of impervious material to prevent**

    a. solid material from blowing away

    b. the lateral expansion of the site

    c. decomposition of the contents

    d. effluent from leaking into the ground beneath it

7.8i **An oxidation ditch is used to**

    a. rear fish for reintroduction

    b. examine a soil profile

    c. treat sewage in small human settlements

    d. store polluted water prior to disposal

7.9i **Which of the following are possible effects of exposure to artificial lighting at night, for example from street lighting?**

    i. Delay in leaf fall in some deciduous trees

    ii. Inhibition of the initiation of pupal diapause in some moths

    iii. Advancement of spring budburst in some trees

    iv. Reduction in the number of flower heads in some plant species

    a. ii, iii and iv

    b. i, ii and iii

    c. i, ii and iv

    d. i, ii, iii and iv

7.10i **Which of the following may be called an 'ecosystem disservice' resulting from urban development?**

    a. Resource depletion

    b. Negative health outcomes

    c. Biodiversity loss

    d. All of the above

7.11i **On a wastewater treatment plant a device that houses microbes that break down sewage sludge and produce methane that may be used as fuel is called**

    a. an anaerobic digester

    b. an activated sludge tank

    c. a sedimentation tank

    d. an aerobic digester

7.12i **A combined sewer system receives wastewater**

    a. from domestic premises and highways only

    b. from roofs and highways only

    c. from domestic and industrial premises, and highways

    d. from domestic and industrial premises only

7.13i **Which of the following urban habitats England has the highest species richness?**

    a. Derelict ground

    b. Allotment gardens

    c. Public parks

    d. Private gardens

7.14i **The application of sewage sludge to agricultural land must be carefully controlled and monitored because**

    a. it may contaminate grazing land and compromise milk quality and animal health

    b. it may contaminate fruit crops with human pathogens

    c. it may contain heavy metals that could accumulate in food chains

    d. All of the above

7.15i **In the 1970s water entering the sewerage system showed an increase in lead levels during storms. This was most likely to have originated from**

    a. fuel spillages on roads

    b. paint

    c. industrial cleaning agents

    d. factories

7.16i **Sewage produced in industrial areas is likely to contain PTEs. The abbreviation PTE refers to**

    a. a chemical produced in the manufacture of plastics

    b. a class of organic solvents

    c. a by-product of the breakdown of commonly used cleaning agents

    d. a number of chemicals that may be toxic in sufficiently high concentrations

7.17i **The initial runoff is often the most polluted volume of water resulting from a rain storm and is called the**

    a. first rush

    b. leading surge

    c. first flush

    d. initial flow

7.18i **The diagram below (Fig. 7.3) is a simple representation of the concept of**

    a. urban metabolism

    b. urban assimilation

    c. urban anabolism

    d. urban catabolism

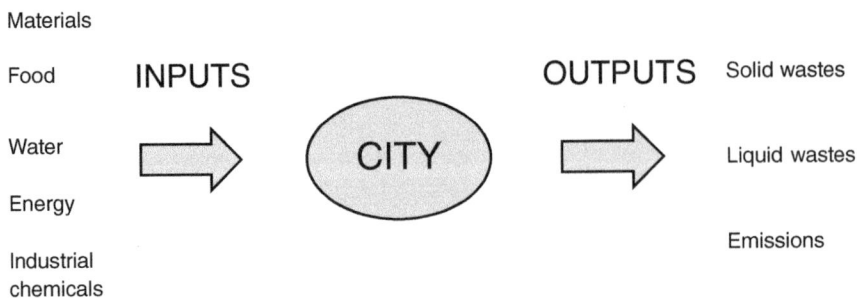

Fig. 7.3.

7.19i   A study of striped field mice (*Apodemus agrarius*) in urban and rural habitats found that urban dwellers were bolder, more explorative and more flexible in some traits than their rural conspecifics. It was concluded that the ability of individual animals to cope with human-induced rapid environmental change (HIREC) was determined by their

  a. intelligence

  b. personality

  c. intuition

  d. instincts

7.20i   When the groundwater level is low sewage may leak into it from a leaking sewer. This movement of water is known as

  a. exclusion

  b. elimination

  c. exfiltration

  d. ejection

Advanced

7.1a   The term 'urban ecology' has been attributed to Robert E. Park of the Chicago School of

  a. Biology

  b. Ecology

  c. Geography

  d. Sociology

7.2a   Brownfield sites are often covered in brick rubble. Rubble soils consisting primarily of brick tend to be deficient in

  a. magnesium

  b. nitrogen

  c. potassium

  d. phosphorus

7.3a   Which of the following data (Table 7.1) represent the relation-
        ship between the percentage lichen cover on trees and distance
        from the city of Belfast in Northern Ireland in the 1960s?

Table 7.1

Distance from Belfast (miles)	Percentage lichen cover on trees			
	A	B	C	D
3	73	3	3	48
4	72	18	18	57
6	57	30	30	72
7	48	73	48	73
9	30	72	57	30
10	18	57	72	18
12	3	48	73	3

    a. A

    b. B

    c. C

    d. D

7.4a   Birds sing louder in noisy environments than in quiet
        environments. This is an example of the

    a. Lumley effect

    b. Lansdowne effect

    c. Lindley effect

    d. Lombard effect

7.5a   Sewage produced in urban areas should be treated before
        it is released into rivers or the sea. In what sequence
        does untreated sewage pass through the components of
        a wastewater treatment plant using the activated sludge
        process?

    a. inlet → screen → grit chamber → primary settlement tanks →
        biological filter → secondary settlement tanks → outlet

    b. inlet → screen → grit chamber → primary settlement tanks →
        aeration tanks → secondary settlement tanks → outlet

    c. inlet → screen → grit chamber → primary settlement tanks → secondary settlement tanks → biological filter → outlet

    d. inlet → primary settlement tanks → grit chamber → screen → aeration tanks → secondary settlement tanks → outlet

7.6a **The purpose of a screen at the inlet in a wastewater treatment plant is to**

    a. remove non-biological solids and other large items from the effluent

    b. prevent fish from entering the treatment plant

    c. reduce the flow of effluent into the plant to prevent flooding

    d. separate large biological solids from liquid effluent

7.7a **The tree whose bark is shown in Fig. 7.4 was widely planted in Britain during the Industrial Revolution and is well known for its capacity to remove pollutants from the atmosphere due to its ability to shed and replace its bark. The tree is**

Fig. 7.4.

a. an English oak (*Quercus robur*)

b. a Manchester poplar (*Populus nigra betulifolia*)

c. a London plane (*Platanus* x *hispanica*)

d. a Norway maple (*Acer platanoides*)

7.8a **The number of species of breeding birds recorded in parks located at different distances from the centre of London is shown in Fig. 7.5, based on data collected by Hounsome (1979). The relationship between these two variables may be described as**

a. a negative correlation

b. a positive correlation

c. no correlation

d. a negative association

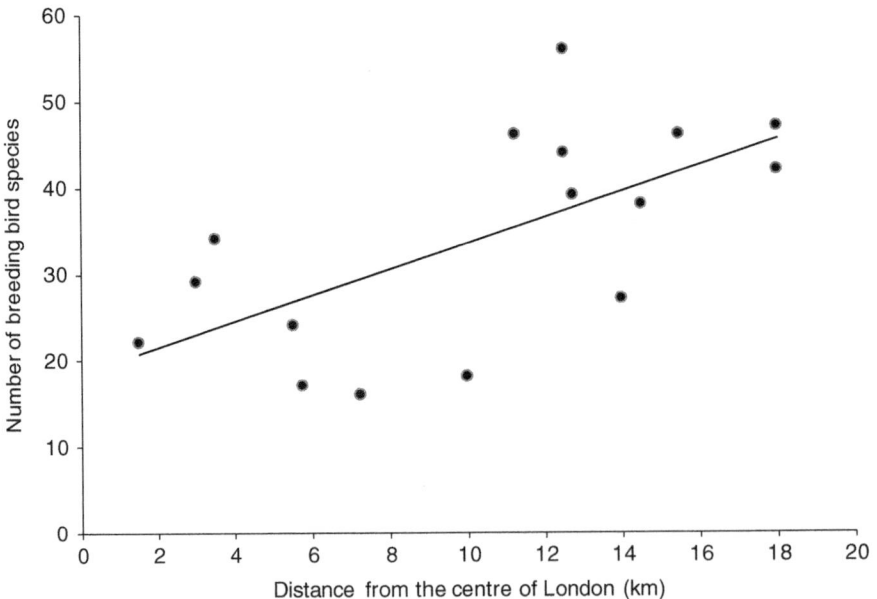

Fig. 7.5.

7.9a The large tank in Fig.7.6 is located in an operational wastewater treatment plant and is normally empty. It is called a

    a. primary settlement tank

    b. storm water retention tank

    c. secondary settlement tank

    d. anaerobic digester

Fig. 7.6.

7.10a Mixed Waste Processing Facilities (MWPFs) accept unsorted waste and recyclable materials. Which of the following options in Table 7.2 (A-D) correctly identifies the differences in performance of MWPFs and Materials Recovery Facilities (MRFs)?

    a. A

    b. B

    c. C

    d. D

Table 7.2

	A		B		C		D	
	MRF	MWPF	MRF	MWPF	MRF	MWPF	MRF	MWPF
Collection cost	Lower	Higher	Lower	Higher	Higher	Lower	Higher	Lower
Capital cost	Lower	Higher	Lower	Higher	Lower	Higher	Lower	Higher
Operating cost	Lower	Higher	Lower	Higher	Lower	Higher	Higher	Lower
Contamination	Higher	Lower	Lower	Higher	Lower	Higher	Lower	Higher
Quality of recovered materials	Higher	Lower	Lower	Higher	Higher	Lower	Higher	Lower
Recovery rates	Higher	Lower	Higher	Lower	Higher	Lower	Lower	Higher
Revenue from recycleable materials	Higher	Lower	Higher	Lower	Higher	Lower	Lower	Higher

7.11a **Table 7.3 shows the main types of urban woodland in Britain. Complete the table by replacing the letters K, L, M and N with terms indicating their origin and degree of naturalness selected from Table 7.4.**

    a. A

    b. B

    c. C

    d. D

Decreasing age of site as woodland
→

		M	N
Increasing naturalness ↑	K	Ancient semi-natural woodland dating from the Middle Ages or earlier	Woodland that has developed through natural colonisation of open sites over the last 4 centuries
	L	Ancient woods clear-felled and replanted with an even-aged stand	$17^{th} - 20^{th}$ century plantations

Table 7.3

Table 7.4

	K	L	M	N
A	Plantation	Secondary	Primary	Semi-natural
B	Semi-natural	Plantation	Primary	Secondary
C	Plantation	Semi-natural	Primary	Secondary
D	Semi-natural	Plantation	Secondary	Primary

7.12a **An area with a circular urban metabolism, compared with one exhibiting a linear urban metabolism, should**

    a. produce less organic waste

    b. recycle more materials

    c. produce lower atmospheric emissions

    d. do all of the above

7.13a The temperature at different points in a cross section of an urban-rural landscape is illustrated in Fig. 7.7. Which of the four graphs (A, B, C or D) illustrates the heat island effect?

    a. A

    b. B

    c. C

    d. D

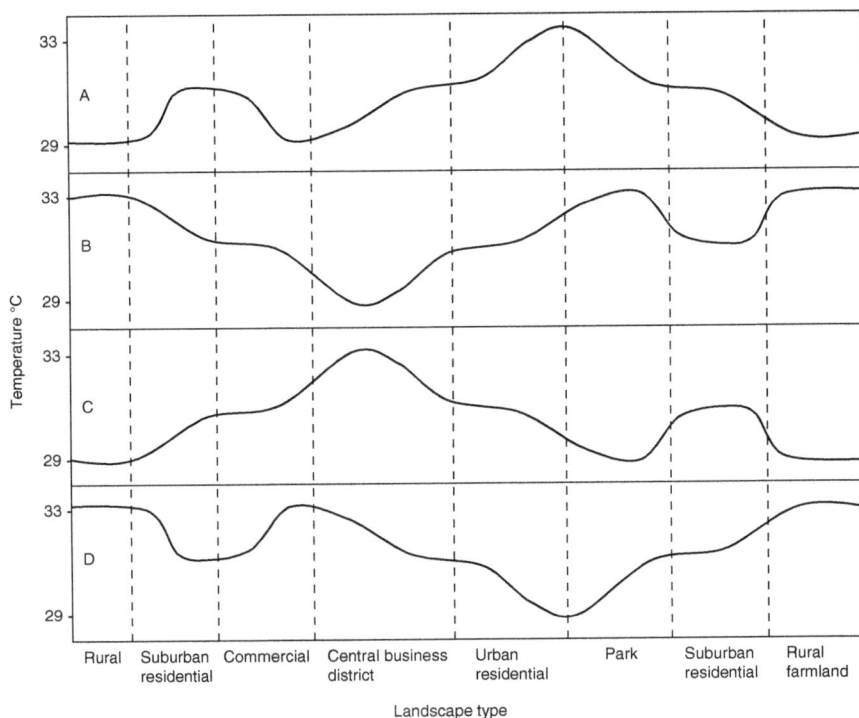

Fig. 7.7.

7.14a A large wastewater treatment works receives a flow of 30,000 litres/second. How many cubic metres does the works treat in 24 hours?

    a. 2.1 million m³

    b. 2.6 million m³

    c. 3.1 million m³

    d. 3.7 million m³

7.15a **Which of the following variables must be quantified to calculate dry weather flow (DWF) to a wastewater treatment plant?**

i.   The population of the catchment (number of people served)

ii. The per capita domestic flow (litres/head/day)

iii. The dry weather infiltration (litres/day)

iv. The trade effluent flow (litres/day)

v.   The runoff from roofs and highways in the catchment during a storm (litres/day)

vi. The ambient temperature

a. i, ii and iii

b. i, ii and iv

c. i, ii, iii and iv

d. i, ii, iii, v and vi

7.16a **In 1887 Ellen Swallow Richards began conducting a sanitary survey of the inland waters of Massachusetts and produced a map of the state showing the extent of manmade pollution. This map is known as the**

a. Normal Chlorine Map

b. Normal Oxygen Map

c. Normal Nitrate Map

d. Normal Ammonia Map

7.17a **Colonies of feral cats (*Felis catus*) are common in urban areas, especially dockyards, the grounds of old hospitals and city centres. Their social organisation resembles that of the**

a. tiger (*Panthera tigris*)

b. lion (*Panthera leo*)

c. leopard (*Panthera pardus*)

d. snow leopard (*Panthera uncia*)

7.18a Studies of birds visiting garden feeders have shown that the species that dominate these feeders are those that are

   a. the smallest

   b. the largest

   c. the fastest

   d. the most intelligent

7.19a The high density of people in urban areas makes them ideal as participants in large-scale ecological studies of species such as garden birds, snails, urban trees and so on. Such studies are often referred to as

   a. citizen science projects

   b. people science projects

   c. public science projects

   d. popular science projects

7.20a A study of ants in Manhattan, New York, found that exotic species were equally common across all habitats (Savage *et al.*, 2014). The authors concluded that

   a. fine-scale heterogeneity in the chronic stress of urban habitats may be an irrelevant structuring force for urban animal communities

   b. coarse-scale heterogeneity in the chronic stress of urban habitats may be an underappreciated, but important structuring force for urban animal communities

   c. fine-scale homogeneity in the chronic stress of urban habitats may be an underappreciated, but important structuring force for urban animal communities

   d. fine-scale heterogeneity in the chronic stress of urban habitats may be an underappreciated, but important structuring force for urban animal communities

8 Global Environmental Change and Biodiversity Loss

This chapter contains questions about global warming, other aspects of environmental change and global biodiversity loss.

Foundation

8.1f The 'greenhouse effect' is largely the result of an excess of what gas in the atmosphere?

a. Ozone

b. Carbon dioxide

c. Nitrogen

d. Carbon monoxide

8.2f Which former US vice-president was featured in the documentary *An Inconvenient Truth* about his campaign to educate people about global warming?

a. Walter Mondale

b. Dan Quayle

c. Dick Cheney

d. Al Gore

8.3f The mass (weight) of carbon dioxide produced by a person over a given period of time is called their

a. carbon output

b. carbon footprint

    c. carbon impact

    d. carbon fingerprint

8.4f **As a result of international concern about acid rain, which of the following was fitted to many coal-powered power stations?**

    a. Carbon capture devices

    b. Sulphur dioxide scrubbers

    c. More efficient cooling towers

    d. Carbon monoxide scrubbers

8.5f **The Living Planet Index is a measure of global biodiversity loss produced by**

    a. the Zoological Society of London (ZSL) and the World Wide Fund for Nature (WWF)

    b. the World Wide Fund for Nature (WWF) and the International Union for the Conservation of Nature (IUCN)

    c. the United Nations Environment Programme (UNEP)

    d. the United Nations Environment Programme (UNEP) and the Zoological Society of London (ZSL)

8.6f **The term 'net zero' generally refers to balancing emissions and absorption of**

    a. sulphur

    b. hydrogen

    c. nitrogen

    d. carbon

8.7f **Which of the following is not a greenhouse gas?**

    a. CH_4

    b. CO_2

    c. O_3

    d. N_2

8.8f Which of the following habitat types is least important in removing carbon from the atmosphere?

   a. Forest

   b. Marshland

   c. Tundra

   d. Temperate grassland

8.9f Which unprecedented event in 2019-2020 caused millions of deaths of animals in Australia?

   a. Extensive bush fires

   b. An epidemic of a fatal virus that affected birds and mammals

   c. An extensive drought

   d. Extensive flooding

8.10f The purpose of adding methane-reducing supplements to cattle feed is to

   a. reduce greenhouse gas emissions from agriculture

   b. increase the efficiency of meat production

   c. improve the welfare of cattle

   d. reduce the nuisance caused by the odour of methane in heavily populated areas near farms

8.11f Which of the following sectors is not a source of greenhouse gas emissions?

   a. Agriculture

   b. Energy supply

   c. Waste management

   d. None – they all produce greenhouse gas emissions

8.12f The International Union for the Conservation of Nature (IUCN) has assessed the risk of extinction for many thousands of species. Of those species assessed, which of the following taxa had the highest percentage of species deemed to be threatened with extinction in 2020?

    a. Mammals

    b. Amphibians

    c. Birds

    d. Reptiles

8.13f Which of the following are characteristics of insect species at the greatest risk of extinction?

    a. Generalist feeders with a single generation per year

    b. Specialist feeders with more than one generation per year

    c. Generalist feeders with more than one generation per year

    d. Specialist feeders with a single generation per year

8.14f A species that was thought to be extinct but has been rediscovered is known as a

    a. Lazarus species

    b. Jacob species

    c. Elijah species

    d. Noah species

8.15f The climate range in which a species currently exists is called its

    a. climate envelope

    b. climate tolerance

    c. climatic capacity

    d. climate capability

8.16f Global sea level is affected by

    a. the water added to the oceans by melting ice sheets and the expansion of seawater as it warms

    b. the expansion of seawater as it warms and tectonic activity

    c. The water added to the oceans by melting ice sheets and tectonic activity

    d. the water added to the oceans by melting ice sheets, the expansion of seawater as it warms and tectonic activity

8.17f **The Living Planet Index (LPI) measures trends in populations of**

a. vertebrate species

b. all animal taxa

c. all animal and plant taxa

d. mammal and bird species

8.18f **Who presented a paper to the Stockholm Physical Society in 1895 entitled 'On the Influence of Carbonic Acid in the Air upon the Temperature of the Ground', an early consideration of global warming?**

a. Anders Celsius

b. Alfred Nobel

c. Svante Arrhenius

d. Christopher Polhem

8.19f **In 2018 the four countries that emitted the most carbon dioxide into the atmosphere are listed in which column of Table 8.1.**

Table 8.1

Highest ranked emitters	A	B	C	D
1st	China	USA	China	China
2nd	USA	China	USA	India
3rd	Germany	India	India	Russian Federation
4th	Japan	Japan	Russian Federation	USA

a. A

b. B

c. C

d. D

8.20f Which of the following is associated with most of the mass extinctions experienced on Earth?

a. An asteroid impact

b. An increase in volcanic activity

c. An increase in sun spot activity

d. Global flooding events

Intermediate

8.1i Dendrochronology is a science that is used to study

a. historical biodiversity loss

b. climate history

c. pollution history

d. fossil history

8.2i The numbers in the diagram below (Fig. 8.1) represent carbon emissions by industry and absorption by forests (in arbitrary units). Which of the diagrams (A-D) illustrates a condition that may be described as carbon neutral?

a. A

b. B

c. C

d. D

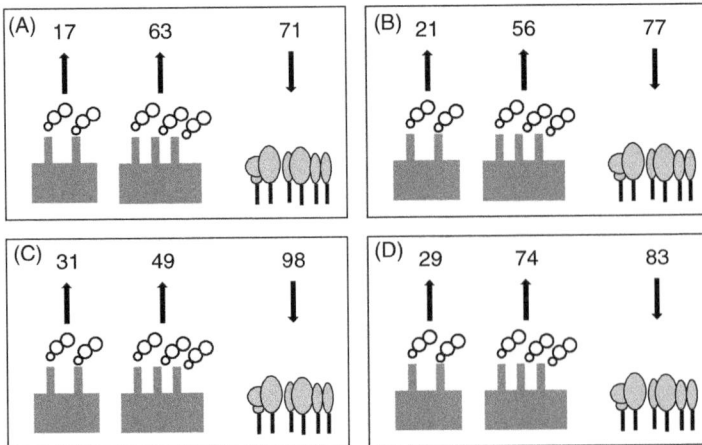

Fig. 8.1.

8.3i   **Which of the following is not likely to be a result of a warming of the climate?**

a. A decline in polar bear numbers due to the breakup of the ice floes on which they travel while hunting for seals

b. A reduction in mountain hare populations in northern Europe due to increased predation resulting from the reduced effectiveness of their white coats as camouflage as snow recedes up mountainsides

c. A deterioration of crop-growing conditions in northern and central Europe

d. An increase in mean sea level

8.4i   **Wetlands provide shore defences, act as flood overspill and dissipate storm energy. Collectively these functions provide an ecosystem service known as**

a. safeguarding

b. buffering

c. shielding

d. cushioning

8.5i   **Select the most appropriate word from the list below to complete the following sentence: 'Forests are important in helping to slow down global climate change because they .... carbon dioxide.'**

a. requisition

b. sequester

c. confiscate

d. impound

8.6i   **Who employed the scientists who discovered the 'hole' in the ozone layer of the atmosphere in 1985?**

a. British Antarctic Survey

b. National Aeronautics and Space Administration

c. National Oceanic and Atmospheric Administration

d. Meteorological Office

8.7i Which of the following types of organisms are most likely to adapt rapidly to environmental change?

a. *K*-selected

b. *r*-selected

c. *L*-selected

d. *p*-selected

8.8i Conflicts between polar bears and humans are increasing because

a. the bears are spending more time on land due to the reduction in the amount of sea ice

b. the bears are scavenging food from human settlements due to reduced hunting opportunities on sea ice

c. it is becoming more difficult for indigenous peoples to predict where the bears will occur because of changes in the distribution of sea ice

d. of all of the reasons above

8.9i The buying and selling of permits to emit carbon dioxide to the atmosphere between industries is known as

a. carbon trading

b. carbon dealing

c. carbon commerce

d. permit trading

8.10i In the 1970s the rise in global sea level was slowed down by

a. a reduction in ice mass loss

b. a reduction in the thermal expansion of the ocean

c. a temporary reduction in global temperature

d. an increase in water impoundment by dams and reservoirs

8.11i The divide between the Cretaceous and Tertiary geological periods during which a mass extinction occurred resulting in the extinction of the dinosaurs is called the

a. C/T boundary

b. T/K boundary

c. T/C boundary

d. K/T boundary

8.12i **The climate cycle in the Pacific Ocean that begins when warm water in the western tropical Pacific moves along the equator eastward towards the South American coast is known as**

a. El Niña

b. El Niño

c. El Toro

d. El Mondo

8.13i **In North America, grizzly bears (*Ursus arctos*) are entering their dens later in the year than normal and emerging from them earlier than normal. The warming of the climate may be contributing to a longer growing season at higher altitude and also in the north of their range than was previously the case. The result of these changes causes**

a. access to plants' roots later in the season, a shorter period of hibernation and a northward extension of their range

b. a shorter period of hibernation and access to food for a longer period

c. an extension of their range further north and a longer hibernation period

d. a shorter period of hibernation and an extension of their range to the south

8.14i **At some level of temperature rise it will become inevitable that a large part of the Antarctic Ice Sheet will melt. The point at which this will occur is known as a**

a. tipping point

b. melting point

c. balance point

d. equilibrium point

8.15i **When the ecological footprint of a region is greater than its biocapacity this indicates a**

    a. biocapacity failure

    b. biocapacity deficit

    c. biocapacity debt

    d. biocapacity loan

8.16i **One cause of extinction in small or fragmented populations is random fluctuations in a species' habitat, such as variations in the weather. This is an example of**

    a. environmental catastrophe

    b. environmental determinism

    c. environmental stochasticity

    d. environmental plasticity

8.17i **A study published by Andermann *et al.* (2020) found that changes in mammalian extinction rates over the previous 126,000 years could best be predicted from**

    a. changes in climate

    b. human population size

    c. the incidence of meteorite strikes

    d. tectonic movements and volcanic activity

8.18i **How many mass extinctions of macroscopic species have occurred on Earth?**

    a. 3

    b. 4

    c. 5

    d. 7

8.19i **Complete the following sentence using one of the numbers listed below: 'In 2018 a United Nations report stated that ....species were threatened with extinction.'**

    a. 100,000

    b. 250,000

c. 500,000

d. 1,000,000

8.20i **Which of the following statements about sea level is false?**

a. Global sea level has been rising since the 19th century

b. Global and local sea levels are measured using the same methods

c. Local sea level is affected by ocean currents, local tectonic activity and other global factors

d. Rising global temperatures cause expansion of water volumes and melt ice

Advanced

8.1a **Ecologically distinct species are defined as those with distinct trait combinations and they are likely to be ecologically irreplaceable. These animals have the most distinct ecological strategies. High ecological distinctiveness in birds and mammals is associated with**

a. high extinction rates or successful hyper-generalism

b. low extinction rates or successful hyper-generalism

c. high extinction rates or successful hyper-specialism

d. low extinction rates or successful hyper-specialism

8.2a **Ocean acidification is the result of**

a. the uptake of nitrogen from the atmosphere

b. the loss of phosphates from fertilisers to the oceans

c. the uptake of carbon dioxide from the atmosphere

d. the uptake of ozone from the atmosphere

8.3a **Atmospheric carbon dioxide concentration at the end of March 2020 was 414ppm. In 1958, when records began, it was 316ppm. What is the percentage increase between the two dates?**

a. 25%

b. 31%

c. 38%

d. 45%

8.4a **Physiognomic analysis of leaves has revealed the relationship between leaf morphology and temperature illustrated in Figs 8.2 and 8.3. This relationship allows us to**

a. predict future changes in rainfall from past relative abundance of leaf type

b. infer paleoclimatic conditions from the relative abundance of leaf types in the fossil record

c. more easily classify living plants from their leaf physiognomy

d. predict the future evolution of leaves from predictions about climate change

Fig. 8.2.

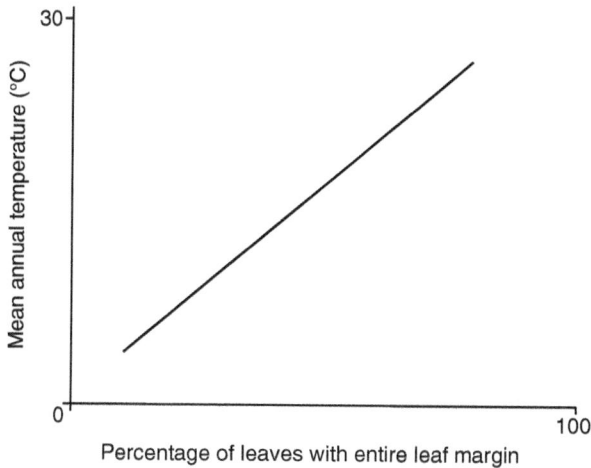

Fig. 8.3

8.5a **Global sea level rose by approximately 230mm between 1880 and 2013. What was the average increase per year?**

    a. 1.13 mm/year

    b. 1.32 mm/year

    c. 1.73 mm/year

    d. 2.16 mm/year

8.6a **Which research station has been monitoring atmospheric carbon dioxide concentrations since 1958?**

    a. The Sphinx Observatory, Switzerland

    b. The Purple Mountain Observatory, China

    c. The Meyer-Womble Observatory, Colorado

    d. The Mauna Loa Observatory, Hawaii

8.7a **Bluetongue is a viral disease that kills sheep, goats and cattle and its recent spread has been linked to climate change. Which of the following statements about this disease is true?**

    a. It is moving northwards through Europe and is transmitted by mosquitoes

    b. It is moving northwards through Europe and is transmitted by midges

    c. It is moving southwards through Europe and is transmitted by mosquitoes

    d. It is moving northwards through Europe and is transmitted by a ticks

8.8a **Changes to the environment that are occurring now will inevitably eventually result in the extinction of certain species in the future. This phenomenon is known as the**

    a. extinction load

    b. extinction liability

    c. extinction debt

    d. extinction burden

8.9a **The allocation of carbon emission permits to industries based on past emissions is known as**

    a. exempting

    b. favouring

    c. grandmothering

    d. grandfathering

8.10a **The first community in the United States to be relocated using federal funding as a result of the consequences of sea level rise was in**

    a. New Orleans, Louisiana

    b. Pascagoula, Mississippi

    c. Isle de Jean Charles, Louisiana

    d. Florida Keys, Florida

8.11a **Complete the following sentence using one of the terms listed below: 'The effect of melting permafrost is to release .... into the atmosphere, thereby accelerating global warming.'**

    i. methane

    ii. carbon dioxide

    iii. sulphur dioxide

    iv. carbon monoxide

a. i only

b. i and ii

c. i and iv

d. ii and iii

8.12a   The five mass faunal extinctions occurred in the

a. Ordivician, Devonian, end of the Permian, Triassic, end of the Cretaceous

b. Cambrian, Devonian, end of the Permian, Triassic, end of the Cretaceous

c. Ordivician, Devonian, end of the Jurassic, Cretaceous, Triassic

d. Cambrian, Cretaceous, Jurassic, Triassic, Devonian

8.13a   What do the species listed in Table 8.2 have in common?

a. They are all island species that have become extinct

b. They have all been saved from extinction by captive breeding programmes

c. They all only exist in zoos

d. They were all rediscovered in the wild having previously been declared extinct

Table 8.2

Vernacular name	Scientific name
Somali sengi	Galageeska revoilii
Bermuda petrel	Pterodroma cahow
Takahē	Porphyrio hochstetteri
Terror skink	Phoboscincus bocourti
'Starry night' harlequin toad	Atelopus arsyecue
Silver-backed chevrotain	Tragulus versicolor

8.14a In August 2020 American scientists working with scientists from the British Antarctic Survey reported the mapping of deep seafloor channels in the Antarctic (Hogan *et al.*, 2020). These provide access to the underside of a glacier for warm water from the deep ocean thereby increasing the rate that it is melting. This glacier was the

    a. Crescent Glacier

    b. Thwaites Glacier

    c. Grigorov Glacier

    d. Leonardo Glacier

8.15a The Living Planet Index (LPI) is a measure of global biodiversity loss. Which of the lines in Fig. 8.4 approximately represents the biodiversity loss between 1970 (the baseline for the index) and 2016?

    a. A

    b. B

    c. C

    d. D

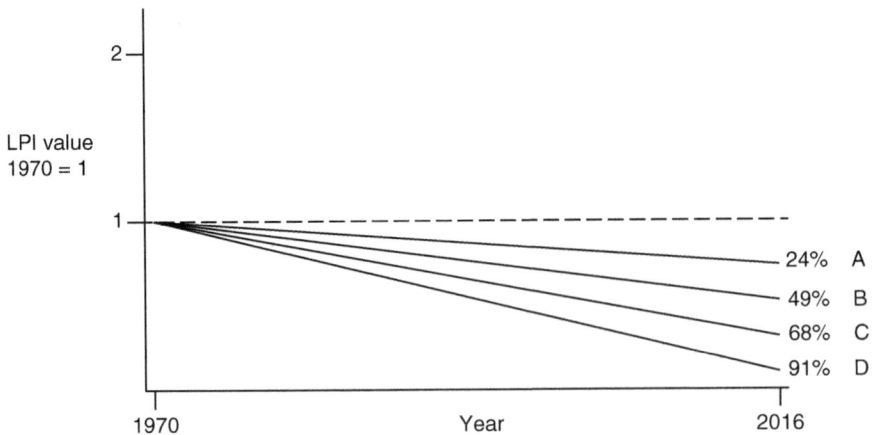

Fig. 8.4.

8.16a **Which of the following is sometimes referred to as the IPCC for biodiversity?**

    a. The United Nations Environment Programme (UNEP)

    b. The International Union for the Conservation of Nature (IUCN)

    c. The Species Survival Commission (SSC)

    d. The Intergovernmental Science-Policy Platform for Biodiversity and Ecosystem Services (IPBES)

8.17a **The Aichi Targets were set to address the underlying causes of**

    a. global climate change

    b. biodiversity loss

    c. habitat loss

    d. atmospheric pollution

8.18a **The impact crater formed by the meteorite that is believed by many scientists to have caused the demise of the dinosaurs and many other species is located in**

    a. the Yucatán Peninsula, Mexico

    b. the Gibson Desert, Western Australia

    c. the Namib Desert, southern Africa

    d. the Kazakh Steppe, Kazakhstan

8.19a **Insert the appropriate word in the sentence below. Spatial modelling of habitat changes predict that a range shift will occur in Asian elephant (*Elephas maximus*) populations due to climate variability resulting from global warming. It suggests that changes in the climatic water balance in monsoon areas will cause elephants to shift their ranges .... along gradients of water availability and drought.**

    a. downwards

    b. upwards

    c. eastwards

    d. westwards

8.20a **Which of the following is not an effect of rising sea levels on coastal communities in Fiji?**

    a. Saltwater intrusion into coastal agricultural land

    b. Beach erosion

    c. Changes in the traditional lifestyle of indigenous people

    d. All of the above are effects of sea level rise

9 Environmental and Wildlife Law and Policy

This chapter contains questions about the laws and policies that protect wildlife and the environment.

Foundation

9.1f Which of the following landscape features is unlikely to be treated as sacred and protected by cultural beliefs within a 'primitive' society?

    a. A mountain

    b. A river

    c. A forest

    d. All of the above are likely to be treated as sacred

9.2f A period of time when the hunting of a particular species is banned by law is called a

    a. closed season

    b. hunting quota

    c. hunting restriction

    d. hunting constraint

9.3f The Convention on International Trade in Endangered Species of Wild Fauna and Flora 1973 (CITES) regulates trade in

    a. specimens of animal species and their products

    b. specimens of plant species and their products

    c. specimens of animal and plant species, living or dead, and their products

    d. specimens of animal and plant species, living or dead

9.4f   A Special Protection Area (SPA) is a type of protected area in the European Union primarily intended to protect

    a. plants

    b. mammals

    c. birds

    d. habitats

9.5f   Natura 2000 is a network of protected areas located in

    a. European countries

    b. Member States of the European Union

    c. African countries

    d. South American countries

9.6f   The Ramsar Convention is primarily concerned with the conservation of

    a. forests

    b. wetlands

    c. savannahs

    d. oceans

9.7f   Only states that have polar bears living wild within their territory are eligible to become parties to the Agreement on the Conservation of Polar Bears 1973. How many states are parties to this Agreement?

    a. 3

    b. 4

    c. 5

    d. 6

9.8f **The headquarters of the United Nations Environment Programme (UNEP) is located in**

    a. Cape Town

    b. New York

    c. Paris

    d. Nairobi

9.9f **A moratorium on commercial whaling was imposed by the International Whaling Commission in the**

    a. 1960s

    b. 1970s

    c. 1980s

    d. 1990s

9.10f **In some legal jurisdictions a drone fitted with a camera may not be used by a hunter on the day hunting is to take place. This is most likely to be because**

    a. the use of a drone may disturb wildlife

    b. such drones may provide live information about the location of game animals and put the hunter at a clear advantage

    c. it would be dangerous for a hunter to use a drone while carrying a firearm

    d. drones may provide unreliable information about the location of wildlife

9.11f **Chemical Z is an insecticide widely sprayed on crops eaten by humans. Prof. Patel believes that this chemical may cause cancer in humans but does not yet have conclusive proof. She believes that the use of chemical Z should be banned. If a government were to implement a ban on this chemical based solely on Prof. Patel's evidence and concerns this would be an application of the**

    a. precautionary principle

    b. safety principle

    c. protective principle

    d. preventative principle

9.12f **Which was the first country to legislate to combat climate change?**

a. New Zealand

b. United Kingdom

c. United States

d. Sweden

9.13f **The Paris Agreement of 2015 was an international agreement concerned with**

a. reducing greenhouse gas emissions

b. reducing biodiversity loss

c. reducing marine pollution

d. reducing desertification

9.14f **The abbreviation IPCC stands for**

a. International Panel on Climate Change

b. Intercontinental Programme for Climate Change

c. Intergovernmental Programme for Climate Change

d. Intergovernmental Panel on Climate Change

9.15f **The Convention on International Trade in Endangered Species of Wild Fauna and Flora 1973 (CITES) regulates trade in wildlife using a system of**

a. import and export permits for species listed in three appendices

b. import permits for species listed in three appendices

c. import and export permits for species listed in two appendices

d. export permits for species listed in three appendices

9.16f **The ICRP is the**

a. International Commission on Radioactive Pollution

b. International Committee on River Pollution

c. International Commission on Radiological Protection

d. International Convention on Radioactive Pollutants

9.17f **The organisation known as TRAFFIC is concerned with monitoring**

    a. trade in wild animals

    b. trade in wild plants

    c. trade in illegal wood products

    d. all of the above

9.18f **Which of the following acronyms refers to an international convention that protects a particular taxon of animals in Europe?**

    a. EUROBEARS

    b. EUROBIRDS

    c. EUROFROGS

    d. EUROBATS

9.19f **Complete the following sentence using a term from the list below: 'An important principle of the environment policy of the European Union is the '...........pays principle.'**

    a. public

    b. polluter

    c. producer

    d. customer

9.20f **Where was the Earth Summit held in 1992?**

    a. Rio, Brazil

    b. Washington, United States

    c. London, United Kingdom

    d. Lima, Peru

Intermediate

9.1i Complete the following sentence using one of the words listed below: 'In 1989 President Daniel arap Moi of Kenya set fire to a large quantity of .... in front of the world's press to publicise the damage done to wildlife by poachers.'

a. rhinoceros horn

b. elephant ivory

c. bushmeat

d. animal skins

9.2i The official forensic laboratory of the Convention on International Trade in Endangered Species of Wild Fauna and Flora 1973 (CITES) is located in

a. Germany

b. United Kingdom

c. United States

d. Switzerland

9.3i The taking of whales (whaling) is permitted by the International Whaling Commission (IWC) for what purpose?

a. Scientific study only

b. Under quotas by indigenous peoples with a long tradition of subsistence whaling only

c. Scientific study and under quotas by indigenous peoples with a long tradition of subsistence whaling

d. Never

9.4i Which of the following laws requires the establishment of Special Areas of Conservation (SACs)?

a. The Ramsar Convention

b. The United Nations Convention on Biological Diversity

c. The European Union Wild Birds Directive

d. The European Union Habitats Directive

9.5i The CAMFIRE project returned the ownership of natural resources, including wildlife, to some rural communities in

a. Zimbabwe

b. Zambia

c. Kenya

d. South Africa

9.6i World Heritage Sites include some places of great ecological importance (e.g. Ngorongoro Crater in Tanzania). Such sites are designated by

a. the United Nations Environment Programme

b. the United Nations Educational, Scientific and Cultural Organization

c. the national government of the country in which they are located

d. the General Assembly of the United Nations

9.7i The international organisation created to address the multiple crises facing humanity and the planet is the

a. Club of Rome

b. Club of Paris

c. Club of Berlin

d. Club of Brussels

9.8i Under CITES, a species is defined as

a. an interbreeding group of animals that is capable of producing viable offspring

b. any animal or plant whether alive or dead

c. any recognisable part of any animal or plant

d. any species, subspecies or geographically separate population thereof

9.9i Which of the following statements is true?

a. The WWF created TRAFFIC to monitor wildlife trade

b. The IUCN created MONITOR to report on wildlife trafficking

    c. The IUCN and WWF created TRAFFIC to monitor wildlife trade

    d. The WWF created MONITOR to report on wildlife trafficking

9.10i **Which of the following is/are not permitted in areas defined by federal law as 'wilderness' in the United States?**

    a. The use of chainsaws

    b. The use of motorised boats

    c. The use of motor cars and trucks

    d. All of the above

9.11i **MARPOL is the acronym used for an international convention concerned with preventing**

    a. pollution from ships

    b. atmospheric pollution

    c. biodiversity loss

    d. land contamination

9.12i **The *World Conservation Strategy* was published in 1980 by**

    a. the International Union for the Conservation of Nature (IUCN), the United Nations Environment Programme (UNEP) and the World Wildlife Fund (WWF)

    b. the International Union for the Conservation of Nature (IUCN) and the World Wildlife Fund (WWF)

    c. the International Union for the Conservation of Nature (IUCN), the United Nations Environment Programme (UNEP) and the United Nations Educational, Scientific and Cultural Organization (UNESCO)

    d. the International Union for the Conservation of Nature (IUCN), the World Wildlife Fund (WWF) and the United Nations Educational, Scientific and Cultural Organization (UNESCO)

9.13i The 'Cod Wars' were a series of confrontations resulting from a dispute over fishing rights in the North Atlantic Ocean between the United Kingdom and

    a. The Netherlands

    b. Norway

    c. Iceland

    d. Denmark

9.14i Which of the following types of trap has been widely banned to protect fur-bearing animals?

    a. Leg-hold traps

    b. Larsen traps

    c. Longworth traps

    d. Sherman traps

9.15i In 2010 an international summit was held in St Petersburg, Russia, to discuss the conservation of

    a. gorillas

    b. tigers

    c. elephants

    d. whales

9.16i The United Nations Convention on the Law of the Sea (UNCLOS) 1982 gave coastal states the exclusive right to manage and exploit fisheries and other marine resources in an Exclusive Economic Zone (EEZ) that extends seaward from the coast to a distance of up to

    a. 100 nautical miles

    b. 150 nautical miles

    c. 200 nautical miles

    d. 250 nautical miles

9.17i   The Basel Convention on the Control of Transboundary Movements of Hazardous Wastes and their Disposal 1989 does not cover the movement of

a. plastic waste

b. recyclable waste

c. heavy metal waste

d. radioactive waste

9.18i   To what does the acronym JARPA apply?

a. The disposal of nuclear waste

b. The control of atmospheric pollution

c. The hunting of whales

d. The control of the wildlife trade

9.19i   Complete the name of the following convention using one of the terms below: 'The United Nations Convention to Combat ....in Those Countries Experiencing Serious Drought and/or .... Particularly in Africa.'

a. Deforestation

b. Desertification

c. Water Pollution

d. Global Warming

9.20i   The Convention on the Conservation of Migratory Species of Wild Animals 1979 is also called the

a. Bonn Convention

b. Washington Convention

c. London Convention

d. Bern Convention

Advanced

9.1a **The purpose of the Montreal Protocol was to reduce**

   a. the discharge of wastes into the oceans

   b. the emission of chlorofluorocarbons (CFCs) to the atmosphere

   c. contamination of land by heavy metals

   d. the use of agrichemicals

9.2a **Which country instituted proceedings in the International Court of Justice against Japan in 2010 for pursuing a large-scale programme of whaling in Antarctica in breach of its obligations under the International Convention for the Regulation of Whaling 1946?**

   a. Australia

   b. United Kingdom

   c. United States

   d. Canada

9.3a **Laundering of wildlife consignments whose movements are regulated by CITES involves**

   a. mislabelling shipments so that they appear to contain species that are not listed in any of the CITES appendices

   b. sending the shipment directly to the country of final destination with incorrect export documents

   c. sending the shipment directly to the country of final destination with incorrect import documents

   d. disguising the true origin of a shipment by sending it to its final destination via a third country that re-exports it with new documentation

9.4a **A legal agreement between a landowner and the government to prevent an area of land from further environmental degradation in exchange for tax benefits is called a**

   a. conservation covenant

   b. environmental covenant

c. conservation contract

d. conservation agreement

9.5a **When part of a country's foreign debt is purchased by a conservation organisation in return for being permitted to establish conservation projects within the debtor nation's territory, this is called a**

a. debt-for-nature-barter

b. debt-for-nature-trade

c. debt-for-nature-switch

d. debt-for-nature-swap

9.6a **Which of the following functions apply to Biosphere Reserves designated as part of UNESCO's Man and the Biosphere Programme?**

a. Conservation and the development of research

b. Ecological monitoring and education

c. Sustainable resource use and logistic support of research

d. All of the above

9.7a **The international agreements that protect the Antarctic are collectively known as the**

a. Antarctic Treaty Complex

b. Antarctic Treaty Collection

c. Antarctic Treaty System

d. Antarctic Treaty Scheme

9.8a **In the European Union, which of the following is not listed as a European Protected Species under the Habitats Directive?**

a. Common otter (*Lutra lutra*)

b. Red squirrel (*Sciurus vulgaris*)

c. Sand lizard (*Lacerta agilis*)

d. Natterjack toad (*Bufo calamita*)

9.9a In which CITES Appendix is the Asian elephant (*Elephas maximus*) listed?

    a. Appendix I

    b. Appendix II

    c. Appendix III

    d. Appendix IV

9.10a Which of the following types of organisms are not listed in the European Union as European Protected Species under the Habitats Directive?

    a. Birds

    b. Fungi

    c. Lichens

    d. None of the above is so listed

9.11a In some jurisdictions the nests of some particular bird species are protected by law even when they are not being used. These tend to be

    a. rare raptors that are site faithful

    b. rare raptors that are not site faithful

    c. rare raptors whether or not they are site faithful

    d. all species that are site faithful

9.12a One way of reducing pressure on popular protected areas and encouraging tourists to visit less popular areas is to adopt a policy of

    a. differential pricing

    b. seasonal pricing

    c. defensive pricing

    d. strategic pricing

9.13a The main aim of the Paris Agreement of 2015 was to

    a. keep global temperature rise this century below 2.5°C

    b. keep global temperature rise this century to well below 2°C above preindustrial levels

    c. pursue efforts to limit temperature increase to 1.5°C

    d. achieve b and c above

9.14a **An international agreement signed by the United States and Canada protects a single migratory herd of**

    a. bison (*Bison bison*)

    b. caribou (*Rangifer tarandus*)

    c. wapiti (*Cervus canadensis*)

    d. moose (*Alces americanus*)

9.15a **Which of the following organisations was established to finance the protection of the global commons and is administered by the World Bank?**

    a. Global Environmental Facility (GEF)

    b. World Wide Fund for Nature (WWF)

    c. International Union for the Conservation of Nature (IUCN)

    d. World Conservation Monitoring Centre (WCMC)

9.16a **The Trail Smelter Case established that no state has the right to use, or permit the use of, its territory in such a manner as to cause injury by fumes in, or to the territory of, another or the properties or persons therein. This case was bought by**

    a. Canada against the United States

    b. the United States against Russia

    c. the United States against Canada

    d. Mexico against the United States

9.17a **The Arusha Conference of 1961 was held to discuss the future of wildlife in**

    a. Africa

    b. United States

    c. India

    d. The Middle East

9.18a **Which of the elements of Fig. 9.1 are not obligations imposed on Parties to the United Nations Convention on Biological Diversity 1992?**

a. 8 and 9

b. 3, 5 and 7

c. 2, 5 and 6

d. All of the elements of Fig. 9.1 describe obligations of the parties

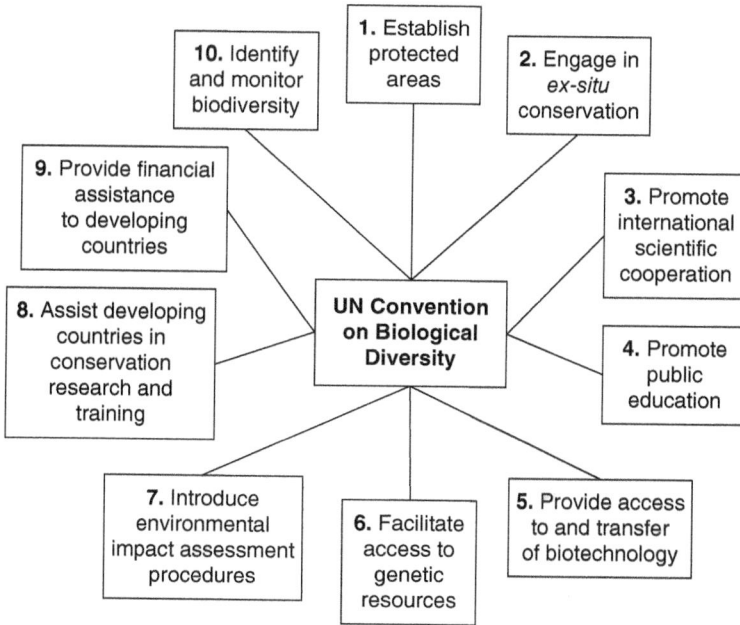

Fig. 9.1.

9.19a **Under international law, the marine living resources of the oceans and seas do not usually include**

a. crustaceans

b. walruses

c. manatees

d. sea otters

9.20a **The European Environment Agency is an agency of the**

    a. Council of Europe

    b. United Nations

    c. European Union

    d. European Commission

10 Environmental Assessment, Monitoring and Modelling

This chapter contains questions about the equipment and methods used to conduct environmental assessments, monitor biodiversity and model ecological systems.

Foundation

10.1f Which of the following is used to trap entire flocks of birds?

    a. A Larsen trap

    b. A cannon net

    c. A Longworth trap

    d. A spring trap

10.2f A biodiversity index measures

    a. the number of types of species present only

    b. the number of individual organisms present only

    c. the number of types of species present and the evenness of the distribution of individuals between these species

    d. the evenness of the distribution of individuals between the species present only

10.3f In forest surveys, tree height can be measured with

    a. an anemometer

    b. a clinometer

c. a hygrometer

d. an altimeter

10.4f The ability of scientists to identify and monitor changes in biodiversity is threatened by a decline in interest among young people in

a. ecology

b. taxonomy

c. genetics

d. anatomy

10.5f An ecologist interested in assessing the diversity of terrestrial invertebrates in urban graveyards might collect samples of these animals using

a. Sherman traps

b. harp traps

c. spring traps

d. pitfall traps

10.6f When planning decisions are considered, members of the public and particular communities may express the desire to not have pollution or new developments in their neighbourhood. This is abbreviated to the acronym

a. DIMBY

b. NIMBY

c. WIMBY

d. RIMBY

10.7f A Heligoland trap is used to capture migratory

a. fishes

b. mammals

c. birds

d. insects

10.8f   Measurements of the quality of ambient air are made

    a. outside

    b. inside buildings

    c. during the night

    d. during windy conditions

10.9f   Environmental impact assessments (EIAs) were first intro-duced in the

    a. 1950s

    b. 1960s

    c. 1970s

    d. 1980s

10.10f Alaga cells were exposed to a range of concentrations of a pesticide to determine its toxicity. This test is an example of a

    a. bioanalysis

    b. bioassay

    c. bioassessment

    d. bioevaluation

10.11f The Berger-Parker index may be used to monitor

    a. biodiversity

    b. particulates in air

    c. soil acidity

    d. the toxicity of a substance

10.12f Which of the following would you expect to produce a digital timestamp?

    a. A camera trap

    b. A temperature data logger

    c. A pop-up satellite archival tag (PSAT)

    d. All of the above

10.13f The acronym SEA stands for

    a. Strategic Environmental Assessment

    b. Special Environmental Assessment

    c. Selective Environmental Assessment

    d. Summative Environmental Assessment

10.14f Photosynthetically active radiation may be measured with a spectrometer and is defined as wavelengths between

    a. 200 and 600nm

    b. 300 and 400nm

    c. 400 and 700nm

    d. 500 and 800nm

10.15f The decibel (dB) scale used to measure noise pollution is

    a. linear

    b. logarithmic

    c. nominal

    d. ordinal

10.16f A portable electronic datalogger could not be used to monitor

    a. soil temperature

    b. insect biodiversity

    c. rainfall

    d. river flow rate

10.17f The Winkler Method is used in water testing to determine the concentration of

    a. dissolved oxygen

    b. nitrates

    c. ammonium

    d. sodium chloride

10.18f The expression below allows the calculation of which property of soil?

$$\frac{Pore\ volume}{Total\ volume}$$

a. Density

b. Bulk density

c. Porosity

d. Texture

10.19f The chloroform fumigation-incubation method is used to measure

a. soil microbial biomass

b. insect respiration

c. the number of parasite eggs in faeces

d. crop plant insect infestations

10.20f An auger is used to sample

a. water

b. air

c. soil

d. plants

Intermediate

10.1i Birds are often used to identify priority areas for conservation because

a. they are found throughout the world in all terrestrial regions and habitats

b. their taxonomy and species identifications are well known

c. there appears to be a good correlation between bird diversity and the diversity of other vertebrates and vascular plants

d. of all of the above

10.2i   A well-studied taxon that is used during ecological surveys as a surrogate for less well-known taxa with which it is associated – perhaps because they have similar habitat requirements – is known as an

a. assessment taxon

b. umbrella taxon

c. index taxon

d. indicator taxon

10.3i   The two most important measures of the performance of a wastewater treatment plant in removing organic matter from sewage are

a. suspended solids and biological oxygen demand

b. biological oxygen demand and chemical oxygen demand

c. suspended solids and settlement rate

d. settlement rate and biological oxygen demand

10.4i   Before a major new development begins (e.g. the construction of a new oil refinery) an Environmental Impact Assessment (EIA) should be completed. The purpose of an EIA is to

a. predict the environmental impacts of the new development once it is completed

b. predict the environmental impacts of the new development during construction and after completion

c. predict the environmental impacts of the new development during construction and after completion, and consider the on-going environmental monitoring that will be needed

d. support the developer's case for locating the development at the proposed site

10.5i   Lead levels in the blood of which of the following organisms have been used as a bioindicator of lead levels in the blood of children in Manhattan, New York?

a. Rats

b. Dogs

c. Pigeons

d. Cats

10.6i **What was the name of the North American experiment conducted in the 1960s that measured the effects of deforestation on the water quality in a river?**

a. The Hubbard Brook study

b. The Howard Brook study

c. The Henry Brook study

d. The Harold Brook study

10.7i **The device in Fig. 10.1 is one of a network of 16 such devices in Britain designed to sample**

a. falling leaves

b. pollen and spores

c. small aerial insects

d. air pollution

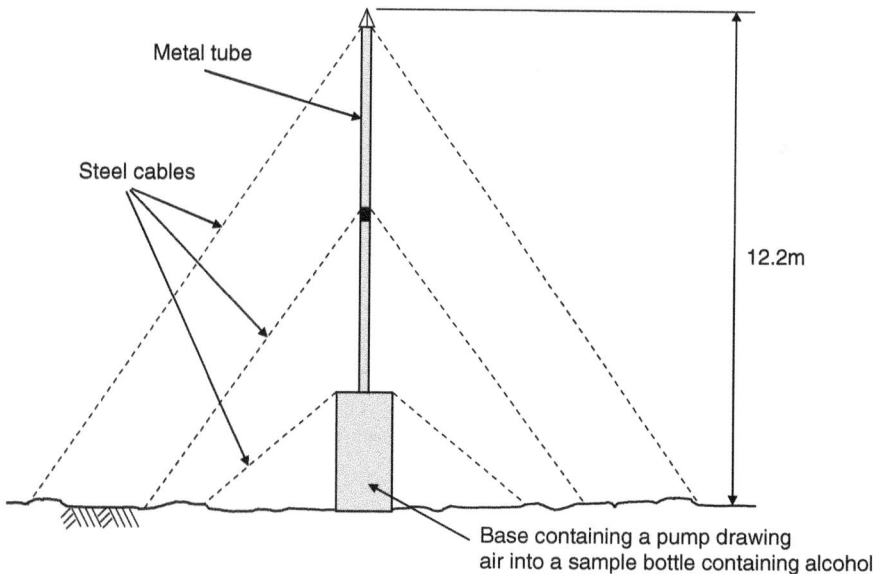

Fig. 10.1.

10.8i **The population size of which of the following species could most easily be determined by the use of camera traps?**

    a. Tigers (*Panthera tigris*) in a forest

    b. Wildebeest (*Connochaetes taurinus*) in a savannah

    c. Starlings (*Sturnus vulgaris*) in a city

    d. Sidewinders (*Crotalus cerastes*) in a desert

10.9i **The quantity of primary production in a savannah ecosystem may be estimated indirectly from measurements of local**

    a. geomorphology

    b. temperature

    c. rainfall

    d. grazing pressure

10.10i **Population growth in a wildlife population may be predicted using a**

    a. Lowton matrix

    b. Leslie matrix

    c. Lawrence matrix

    d. Lowry matrix

10.11i **A wildlife ecologist wishes to simulate the functioning of a woodland ecosystem under a variety of circumstances taking into account the effects of chance events. She should use a**

    a. Monaco simulation

    b. Monte Cassino simulation

    c. Monte Carlo simulation

    d. Montenegro simulation

10.12i **When ecological monitoring of a site occurs over an extended period of time new methods need to be calibrated against old ones to minimize**

    a. methodological conflict

    b. analytical inconsistency

c. methodological drift

d. analytical disparity

10.13i In situations when there is an immediate threat to the environment, for example from development, it may not be possible to make a detailed survey of the species present. Instead, it may be possible to employ simple methods to assess certain indicator groups. Which of the following terms is not used for such an approach?

a. Rapid Biodiversity Assessment (RBA)

b. Rapid Ecological Assessment (REA)

c. Biological Rapid Assessment (BIORAP)

d. Strategic Rapid Assessment (SRAPA)

10.14i Fisheries scientists can gain information about fish populations from samples of scales taken from individuals. Which of the following may be determined from a fish scale?

a. The age of the fish only

b. The growth of the fish in each year only

c. The age and growth of the fish in each year

d. The sex and age of the fish

10.15i A LIDAR scanner produces images using

a. laser light

b. ultrasound

c. radar

d. infrasound

10.16i The assessment of all the environmental impacts associated with all the stages in a product's life from the extraction of raw materials to materials processing, product manufacture, distribution, use and disposal is known as a

a. product impact assessment

b. life cycle assessment

c. sustainability analysis

d. lifespan analysis

10.17i A dispersion model is a method of calculating air pollution concentrations using data about

a. pollutant emissions only

b. atmospheric conditions only

c. pollutant emissions and atmospheric conditions

d. pollutant sources and pollution emissions

10.18i A pyranometer measures

a. soil moisture

b. solar radiation

c. air temperature

d. wind speed

10.19i A modern electronic leaf area meter can be used to measure

a. detached non-perforated leaves only

b. attached non-perforated leaves only

c. attached and detached non-perforated leaves

d. attached and detached perforated and non-perforated leaves

10.20i The fungus *Rhytisma acerinum* causes a common leaf disease in sycamore (*Acer*) trees that manifests itself as 'tar spots'. Figure 10.2 shows the difference in appearance of the leaves between clean air (A) and polluted air (B). A leaf area meter can be used to calculate the number of tar spots/100 cm² of leaf surface (the tar spot index or TSI). In areas of high sulphur dioxide concentrations in the atmosphere

a. the TSI decreases and the spot size increases

b. the TSI increases and the spot size decreases

c. the TSI decreases and the spot size is unaffected

d. the TSI is unaffected and the spot size increases

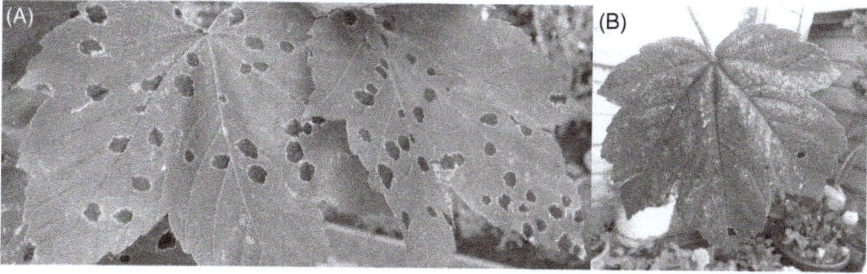

Fig. 10.2.

Advanced

10.1a The effects of distance downstream of a sewage (wastewater) outfall on the abundance of various aquatic organisms are shown in Fig. 10.3. Monitoring of the water quality in a river may be achieved by examining the relative abundance of indicator organisms that require high levels of oxygen and others that can survive in low oxygen conditions. In the table below (Table 10.1) match the lines on the graph with the appropriate taxa.

a. A

b. B

c. C

d. D

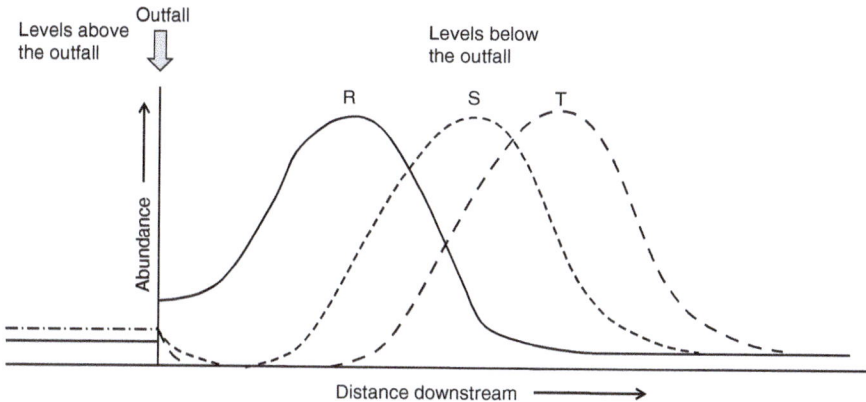

Fig. 10.3.

Table 10.1

Line	A	B	C	D
R	Tubificidae	Chironomus	Asellus	Tubificidae
S	Asellus	Asellus	Tubificidae	Chironomus
T	Chironomus	Tubificidae	Chironomus	Asellus

10.2a **An ecologist measured the invertebrate biodiversity of pond A using Simpson's index and the invertebrate biodiversity of pond B using Menhinick's index. The values obtained were 0.72 for pond A and 0.61 for pond B. Which of the following statements is true?**

a. The biodiversity of invertebrates is higher in pond A than in pond B

b. The biodiversity of invertebrates is higher in pond B than in pond A

c. The two values cannot be directly compared

d. Simpson's index should not be used to measure invertebrate biodiversity

10.3a **The formula below is known as**

a. Simpson's index

b. the Shannon-Weiner index

c. the Lincoln index

d. the Berger-Parker index

$$H = -\Sigma p_i \log_2 p_i$$

Where,

p_i = the decimal fraction of individuals in the ith species.

10.4a **The biocapacity of an ecosystem is an estimate of its production of certain biological materials and absorption and filtering of other materials (e.g. atmospheric carbon dioxide). It is measured as**

a. global hectares/person

b. global hectares/1000 persons

  c. national hectares/person

  d. tonnes/person

10.5a **The difference between the species richness of samples from different sites is known as the**

  a. alpha diversity

  b. beta diversity

  c. theta diversity

  d. sigma diversity

10.6a **The aircraft shown in Fig. 10.4 is being used to count giraffes in a national park in Africa by flying in a straight line of length L (from X to Y) and counting animals within a distance of w on either side. This is known as a strip transect. From the area being sampled, if G = the number of giraffes recorded, the density of giraffes may be calculated using the formula**

  a. $\dfrac{G}{wL}$

  b. $\dfrac{G}{2wL}$

  c. $\dfrac{2wL}{G}$

  d. $\dfrac{2G}{wL}$

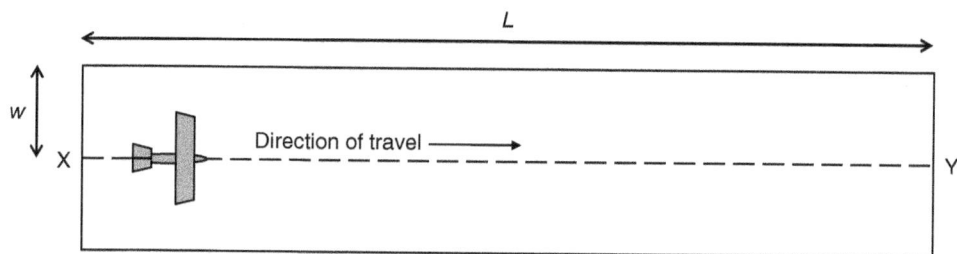

Fig. 10.4.

10.7a   Which of the following techniques could be used to identify future protected areas by mapping the ranges of indicator species, predicting potential ranges from habitat maps and comparing these with the existing protected area network?

   a. Space analysis

   b. Range analysis

   c. Gap analysis

   d. Disparity analysis

10.8a   Complete the following sentence using one of the options (a-d) below: 'The .... model of Laurance and Yensen (1991) predicts the amount of edge-affected area and remaining unmodified core area in isolated habitat fragments.'

   a. core area

   b. edge area

   c. edge area:unmodified area ratio

   d. isolated fragment area

10.9a   Each time a stochastic model of an ecological process is run it always produces

   a. the same prediction

   b. a different prediction

   c. an accurate prediction

   d. an unreliable prediction

10.10a Quantitative models may assist in generating effective conservation actions if

   a. they focus on accurately modelling ecological processes

   b. modellers consider all possible variables

   c. modellers conceptualise their models with the ultimate aim of informing conservation management

   d. extremely powerful computers are used

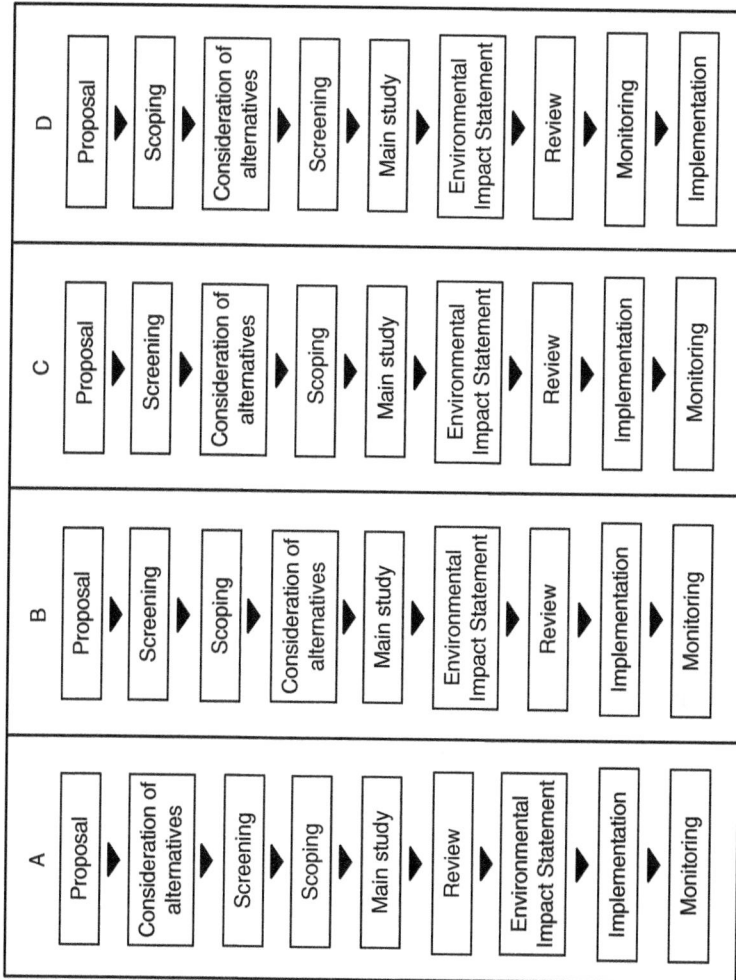

Fig. 10.5.

10.11a Which of the sequences in Fig 10.5 best describes the main stages in an environmental impact assessment (EIA)?

   a. A

   b. B

   c. C

   d. D

10.12a The Braun-Blanquet scale is used to measure

   a. plant cover

   b. biodiversity

   c. air pollution

   d. chemical toxicity

10.13a The growth of an animal population may be predicted using matrix algebra. The expression below relates to a theoretical population of just three generations (0, 1 and 2). The number of offspring born to an individual in the oldest age group is given by

   a. p_1

   b. n_{2t}

   c. n_{2t+1}

   d. f_2

$$\begin{bmatrix} n_{0t+1} \\ n_{1t+1} \\ n_{2t+1} \end{bmatrix} = \begin{bmatrix} f_0 & f_1 & f_2 \\ p_0 & 0 & 0 \\ 0 & p_1 & 0 \end{bmatrix} \times \begin{bmatrix} n_{0t} \\ n_{1t} \\ n_{2t} \end{bmatrix}$$

10.14a Life tables are used to study survivorship in populations. In a life table the number of organisms surviving to age x is shown in a column labelled

   a. l_x

   b. m_x

   c. p_x

   d. q_x

10.15a Table 10.2 shows the species present at two locations, site X and site Y (indicated by √). Calculate the beta diversity index for these sites using the formula below.

$$\text{Beta diversity index} = \frac{2c}{s1 + s2}$$

where,

$s1$ = number of species recorded at site 1

$s2$ = number of species recorded at site 2

c = number of species in common (i.e. recorded at both sites 1 and 2).

The value of the index is

a. 0.37

b. 0.45

c. 0.59

d. 0.61

Table 10.2

Species	Site X	Site Y
A	√	√
B	√	√
C	√	
D		√
E	√	
F	√	
G	√	√
H		√
I	√	
J		√
K	√	√
L	√	√

10.16a Benkwitt (2015) manipulated juvenile Pacific red lionfish (*Pterois volitans*) densities on small patch reefs in the Bahamas to predict the effect of this invasive species on native coral reef fish communities. Some of her results are shown in Fig.10.6. The relationship between lionfish density and the proportional change in the abundance of prey-sized native fishes may be described as

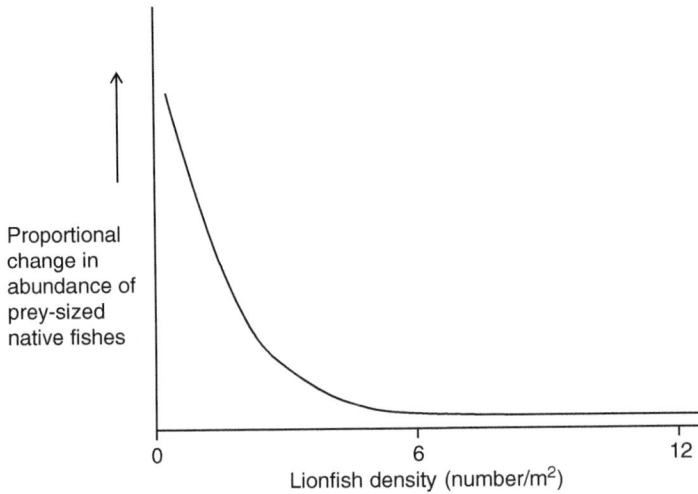

Fig. 10.6.

    a. linear

    b. exponential

    c. negative exponential

    d. logistic

10.17a A cluster analysis could be used to examine

    a. the relationship between the sulphur dioxide concentration in the air at increasing distances from the centre of a city and the number of lichen species recorded

    b. the similarities between soil samples taken from 25 different sites in England

    c. the relationship between light intensity and growth in barley seedlings

    d. the effect of prey density on lion density in Ngorongoro Crater, Tanzania

10.18a A biologist wants to understand the relationship between three different application rates of fertiliser (1, 2 and 3) on plant growth in two species (A and B). The growth values are indicated in Table 10.3 by the values k to p. She should analyse her results using a

a. one-way analysis of variance (one-way ANOVA)

b. two-way analysis of variance (two-way ANOVA)

c. 2 x 2 contingency table

d. correlation coefficient

Table 10.3

	Fertiliser application rate		
Plant Species	1	2	3
A	k	l	m
B	n	o	p

10.19a Which of the following methods would be most appropriate to study how people perceive the threats to their livelihoods from local wildlife species, for example by comparing farmers' perceptions of species causing crop damage with measured crop losses?

a. Game theory

b. Discourse analysis

c. Network analysis

d. Participatory risk mapping

10.20a Which of the following techniques and equipment can be used to measure leaf area index (LAI) in a forest?

a. A plant canopy analyser

b. Ceptometry

c. Hemispherical photography

d. All of the above

11 Answers

A multiple choice question has a stem (the 'question'), a key (the 'answer') and a number of distracters (wrong answers intended to distract the student from the key). This part of the book contains the key to each question along with a brief explanation of why this is correct and, in some cases, what the distracters mean.

Chapter 1 History and Foundations of Applied Ecology and Conservation

1.1f	B	The German zoologist Haeckel first used this term.
1.2f	C	The term was first used by Walter G. Rosen in 1985.
1.3f	D	Yellowstone National Park founded in 1872.
1.4f	B	A grove is an area of woodland.
1.5f	A	The WWF originally focussed on the big game animals of East Africa as this region had suffered significant losses due to commercial hunting and poaching.
1.6f	D	The Greek root *bio-* relates to 'life' and *phil-* relates to 'love'.
1.7f	B	This is a photograph of a model of a dodo. The other species are also extinct birds.
1.8f	D	Ralph Nader was a well-known advocate for consumers, especially in relation to motor vehicle safety.
1.9f	B	Areas classified as 'wilderness areas' are highly valued because of the virtual absence of human activity. The other terms are also categories of protected area used by the IUCN.

1.10f	C	Greta Thunberg campaigned to focus the attention of the public and world leaders on climate change, initially by organising 'strikes' by students in Swedish schools. In 2019 she addressed world leaders at the United Nations Climate Action Summit.
1.11f	B	Norman Myers is correct. Ehrlich and Leopold were advocates for the environment but Sagan was an astronomer.
1.12f	A	Sustainable use of resources allows for their use in a managed fashion so that they are not over-exploited, rather than banning their use altogether.
1.13f	B	*Silent Spring* alerted the world, especially the American public, to the damage being done to ecosystems by chemicals, especially those used in agriculture, such as DDT. The name of the book is a reference to the absence of birdsong.
1.14f	C	The Green Revolution was a response to the need to produce more food for a rapidly growing human population.
1.15f	D	Cultural, social and political ecology are all relatively new branches of ecology that focus on human society and its relationship with the environment.
1.16f	A	Muir was a Scottish-American naturalist who advocated the preservation of wilderness areas in the United States in the 19th century.
1.17f	A	Scott was a British naturalist, writer, artist and broadcaster with a particular interest in wildfowl.
1.18f	B	The Greek root *anthrop-* refers to humans.
1.19f	D	This is *Arctic Sunrise*, a ship operated by Greenpeace, moored in the harbour in Reykjavík, Iceland.
1.20f	A	She established the Green Belt Movement in Kenya in 1977 that organised the planting of large numbers of trees in Kenya.
1.1i	C	The Greek root *ethn-* refers to a people or tribe and botany is the study of plants.
1.2i	D	The Peak District National Park in Derbyshire, England, was established in 1951.
1.3i	A	Thoreau was an American writer who wrote about natural history and the environment, among many other things.
1.4i	C	The Bureau was created to identify insect pests and disseminate information about them.
1.5i	D	A number of buffalo jumps are now protected as important cultural and archaeological sites, e.g. the Head-Smashed-In Buffalo Jump in Alberta, Canada, is a World Heritage Site.
1.6i	B	The dust bowl was the result of soil erosion caused by poor land management by farmers.

1.7i	B	This event brought the issue of air pollution to the attention of the government and led to the passing of the Clean Air Act in 1956.
1.8i	C	*Oryx* is correct. The journal is subtitled 'The International Journal of Conservation'. The distracters are all other antelope species. The oryx is the emblem of Fauna and Flora International.
1.9i	A	The Red List is an inventory of organisms grouped into a number of threat categories produced by the IUCN.
1.10i	A	Greenpeace mounted long-running campaigns against whaling and nuclear power, that included interfering with whaling operations at sea.
1.11i	B	The Sierra Club was founded in San Francisco in 1892. The location and dates in the distracters have no particular significance in this context.
1.12i	C	Passenger pigeons had previously been extremely common in North America and were hunted to extinction. The species is often used as an example of how easy it is to extirpate a species. The last Carolina Parakeet died in Cincinnati Zoo in 1918.
1.13i	B	This is celebrated to mark the anniversary of the birth of the modern environmental movement and was first held in 1970.
1.14i	D	Paul Ralph Ehrlich was Professor of Population Studies at Stanford University, California.
1.15i	A	Conservation psychology is a new branch of science concerned with the interrelationships between human behaviour and conservation.
1.16i	B	Australia has a large number of endemic species (those that occur nowhere else) because it has been isolated from other land masses for a very long period.
1.17i	D	The extensive tropical forests of South America (especially Peru) support large numbers of butterfly species.
1.18i	C	The 'tragedy of the commons' is correct. The distracters are fictitious.
1.19i	D	*Ex-situ* (off-site) conservation projects take place outside the habitat where a species naturally occurs, e.g. in a zoo.
1.20i	B	Barry Commoner was an American biologist. The distracters are all well-known American ecologists.
1.1a	C	Deep ecology is correct. The distracters are fictitious in this context.
1.2a	D	The expression 'tree hugger' is derived from the name of this movement and used as a derogatory term for someone who wants to protect the environment.
1.3a	B	The distracters are the same people linked to incorrect campaigns.
1.4a	A	The distracters are fictitious.
1.5a	B	*The Limits to Growth* was commissioned by the Club of Rome and argued that the world could not sustain its high rates of economic and population growth beyond another 100 years.

1.6a	A	The distracters are other important scientists.
1.7a	A	President Theodore Roosevelt was well known for his interest in hunting and wildlife.
1.8a	B	This was an unofficial report commissioned by the United Nations following the UN Conference of the Human Environment in 1972.
1.9a	B	1798 is correct. The other dates have no significance in this context.
1.10a	D	Paul Müller discovered the insecticidal properties of DDT. The distracters are other insecticides.
1.11a	D	Northeast Greenland National Park is the largest national park and protects almost one million square kilometres.
1.12a	B	Flood protection is concerned with regulating water levels; forest provides food; nature provides spiritual experiences linked to cultures.
1.13a	A	Muir founded the Sierra Club in San Francisco in 1892 to campaign to protect the Earth's natural resources.
1.14a	C	This concept was first developed by Norman Myers. The distracters are other ecologists.
1.15a	B	Durrell trains conservationists from all over the world at its facilities in Jersey, especially those from developing countries.
1.16a	D	Favourableness is correct. The distracters have no meaning in this context but have related meanings.
1.17a	C	The RSPB, under an earlier name, was established in Manchester, England, to campaign against the use of feathers in women's fashion, especially hats.
1.18a	C	The goal (v) is a matter of judgement (policy) and therefore not the role of a manager. The others are technical decisions.
1.19a	B	'Wilderness' is an American concept derived from the vast amount of uninhabited land in North America.
1.20a	A	This became the Fauna Preservation Society and later Fauna and Flora International.

Chapter 2 Environmental Pollution and Perturbations

2.1f	B	Eutrophication in some circumstances is natural. When the process is accelerated by human activity it is referred to as cultural eutrophication.
2.2f	C	The process involves cutting down the natural vegetation and then burning the dead debris, hence 'slash-and-burn'.
2.3f	B	Plastic products enter rivers along their length and then travel to the sea.
2.4f	A	Fast breeder reactors produce fissile material.

2.5f	A	Chemical fertiliser, milk and sewage all contain large amounts of nutrients causing overgrowth of plants. Cadmium waste would be toxic but would not supply nutrients.
2.6f	C	Acid rain is defined as rain with a pH of less than 5.6.
2.7f	A	Chlorofluorocarbons (CFCs) caused damaged to the ozone layer and created a 'hole'.
2.8f	D	Many gunshot pellets contain lead. Lead used to be a constituent of many paints and was used as an additive in petrol to improve engine performance and prevent engine damage.
2.9f	B	Heavy metal-tolerant plants have evolved through the process of natural selection to tolerate levels of metals such as zinc, lead and copper that are toxic to other plants.
2.10f	D	The *Herald of Free Enterprise* was a roll-on/roll-off ferry that capsized in 1987 when the bow doors were left open.
2.11f	A	Potassium is not a heavy metal.
2.12f	D	Mercury poisoning causes a wide range of neurological signs.
2.13f	C	Each organism in a food chain generally consumes large numbers of the organisms in the level below and pollutants taken up by the organisms at the bottom of the food chain become concentrated as a consequence.
2.14f	A	Stomata are small pores in plant leaves through which gases (including pollutants) are exchanged. The distracters are the names of different types of plant tissues.
2.15f	C	A forest fire may cover an extensive (and extending) area while the distracters are limited to a specific point (even if moving).
2.16f	B	Forest is cleared for cattle in South America and palm oil in Southeast Asia.
2.17f	C	Fracking involves pumping water, sand and chemicals into the ground at high pressure to release gas. Methane and carbon dioxide leak from fracking wells as a result.
2.18f	A	PCBs are persistent organic pollutants that are destroyed by incineration and can pass through the placenta.
2.19f	C	Although the loss of tropical forest is widely publicised, wetlands have lost a higher proportion of their total area since 1700.
2.20f	D	The three distracters are primary pollutants, i.e. they are emitted directly from a source. Ozone is formed in the atmosphere as a result of a complex series of chemical reactions that involves sunlight so it is a secondary pollutant.
2.1i	D	The pills contain sex hormones.
2.2i	A	Bhopal was the site of an explosion caused by a gas leak at the Union Carbide India Limited pesticide plant that enveloped the city in poisonous gases in 1984.

2.3i	D	Overgrazing, vegetation clearance and deforestation all cause the loss of soil.
2.4i	C	These forests were greatly affected as circulating air brought acid rain from industry in Western Europe.
2.5i	C	Mercury is a dangerous neurotoxin. It was released from industries into the sea near Minamata and absorbed by fishes and shellfish eaten by local people who began showing signs of mercury poisoning in 1956.
2.6i	B	The Aral Sea is in Central Asia.
2.7i	C	These dams are located on the Yangtze River in China.
2.8i	B	Carbon monoxide combines with haemoglobin in the blood and prevents it from carrying oxygen.
2.9i	A	Predatory fishes are at the top of the food chain so will accumulate the highest concentrations of mercury.
2.10i	D	Nutrients are washed out of soils when trees are removed.
2.11i	B	Detergents may do more harm than good. Oil will biodegrade naturally.
2.12i	B	The testing of nuclear bombs released strontium-90 into the atmosphere. This contaminated agricultural land, was taken up by the grasses consumed by cattle and then passed into their milk.
2.13i	A	Cetaceans use echolocation (sound) to navigate.
2.14i	B	Studies of eggs held by museums showed that their thickness increased after DDT use declined in Britain.
2.15i	D	Dioxins enter the body via the food chain and have a natural affinity for fatty tissue.
2.16i	D	Half-lives are: iodine-131 = 8 days; plutonium-239 = 24,110 years; strontium-90 = 28.8 years; uranium-238 = 4.5 billion years.
2.17i	C	Strontium-90 contaminated the soil, was absorbed by grasses and then eaten by sheep.
2.18i	A	The distracters all contain metals that do not occur in catalytic converters.
2.19i	D	$PM_{2.5}$ refers to particulate matter (PM) with a diameter of less than $2.5\mu m$.
2.20i	A	This volume of water experiences little mixing and has been exposed to nutrient inputs so has become eutrophic, causing an excessive growth of plants.
2.1a	C	Microplastics are pieces of plastics that have a length of less than 5mm as defined by the US National Oceanic and Atmospheric Administration (NOAA).
2.2a	A	A fitness or adaptive landscape can be used to visualise the relationship between different genotypes and success, usually in three dimensions. Fig. 2.3 is a two-dimensional representation of this.

2.3a	A	Transuranic elements have an atomic number greater than that of uranium (i.e. 92).
2.4a	D	Zooxanthellae are single-celled dinoflagellates that live symbiotically with coral and provide the coral polyps with nutrients produced by photosynthesis.
2.5a	A	The Indian Government banned the use of diclofenac in 2006.
2.6a	B	Cold air is trapped under warm air in a temperature inversion. In this situation temperature initially rises with height and air pollution is trapped in the cold air.
2.7a	C	This was a major poisoning incident that occurred in a neighbourhood of Niagara Falls where a community had been built on toxic industrial waste.
2.8a	B	Neonicotinoids are implicated in the decline in pollinating insects, especially bees. They are systemic insecticides and can be present in nectar and pollen.
2.9a	A	DDT breaks down to DDD and DDE. DDC and DDF are fictitious in this context.
2.10a	B	DDT is a chlorinated hydrocarbon.
2.11a	B	Chlorofluorocarbons (CFCs) are broken down by UV light in the stratosphere and the chlorine released destroys ozone.
2.12a	C	Aquatic molluscs use calcium carbonate to produce a hard shell.
2.13a	B	DDT has travelled extensively in the atmosphere and has been detected in parts of the world where it has never been used.
2.14a	C	Cold air trapped underneath warm air prevents the dispersal of pollutants.
2.15a	A	Teratogens are substances that cause birth defects in the embryo or foetus after the mother is exposed to them; carcinogenic means cancer-causing; mutagenic means mutation-causing; ontogenetic means relating to ontogeny (development).
2.16a	B	Nonthreshold pollutants are dangerous at any level. Many substances are harmless (or even beneficial in low concentrations) but dangerous at high levels (e.g. vitamin A).
2.17a	D	A dense gas plume would sink because it would be heavier than the ambient air. Passive and neutral plumes are the same thing and have a temperature similar to the ambient air.
2.18a	A	The impacts of celestial bodies on the Earth have caused extensive damage compared with the relatively localised effects of earthquakes, flood and forest fires.
2.19a	C	PANs are created in the atmosphere and are components of photochemical smog.
2.20a	D	This was an American study that examined the milk teeth of many thousands of young children to determine the extent of exposure to radioactivity from atomic bomb testing.

Chapter 3 Wildlife and Conservation Biology

3.1f	D	Mountain gorillas are not bred in captivity as part of any coordinated captive breeding programme but are currently recovering well in the wild.
3.2f	C	The protection of umbrella species (e.g. tigers) indirectly protects other species in the same community and habitat.
3.3f	A	This starfish gained notoriety because of the destruction it has caused to the Great Barrier Reef as it feeds on coral.
3.4f	C	Such connecting strips of land are usually called wildlife corridors. They allow animals to travel and disperse from one protected area to another.
3.5f	B	Orangutan habitat is being destroyed and fragmented by the removal of forest to create large palm oil plantations.
3.6f	A	Ecologically-based tourism does not have a destructive effect on natural systems. All of the distracters involve permanently removing animals or plants from their natural habitats.
3.7f	D	The distracters are fictitious terms based around the concept of a flag or poster.
3.8f	C	Coniferous plant species are common in northern latitudes but rare in the tropics. The reverse is true for the distracters.
3.9f	B	The Association of Zoos and Aquariums operates in North America. The distracters are other zoo associations that cooperate in other breeding programmes.
3.10f	B	Seeds are stored in a seed bank; frozen sperm and ova are stored in a frozen zoo.
3.11f	C	Fortress conservation involves seizing land from indigenous peoples and infringing their land and human rights. This is an outdated approach to conservation as it should involve local people rather than exclude them.
3.12f	D	These animals were taken to a place where they had not previously existed so this was an introduction.
3.13f	C	Conservation projects that take place in the wild are 'in-situ'.
3.14f	D	Elephants have been controlled by shooting (e.g. in Uganda and South Africa), given contraceptives (South Africa) and moved from one protected area to another (e.g. Kenya).
3.15f	A	Flyways are flight paths used by migrating birds, e.g. the West Pacific Flyway, the Mississippi Flyway.
3.16f	D	This migration occurs between the Serengeti and the Maasai Mara every year as the wildebeest move between Tanzania and Kenya.

3.17f	D	Captive populations are often called insurance populations because, if they are self-sustaining, they can prevent the extinction of a species and may provide individuals for reintroduction to the wild.
3.18f	B	The term is restricted to animals of large body size that exist in a particular area or existed in a particular geological period.
3.19f	A	Studbooks record all of the individuals in a breeding programme along with data on their origin, parents, mates, offspring and other demographic data.
3.20f	B	Breeding animals in a zoo is an example of *ex-situ* conservation. All the distracters refer to *in-situ* projects (i.e. they occur in the wild).
3.1i	A	The Species Survival Commission (SSC) is a network of around 9000 volunteer experts that provide the International Union for the Conservation of Nature (IUCN) with information about biodiversity conservation.
3.2i	C	A demographically extinct population is one whose age structure and reproductive performance make it inevitable that it will not survive.
3.3i	C	The definition of a biodiversity hotspot focuses on endemic plants and areas where there has been a significant loss of natural vegetation. Conservation International listed 36 such areas in 2020.
3.4i	D	Tasmania is not recognised by Conservation International as a biodiversity hotspot.
3.5i	B	In the past developed countries have discovered new drugs as a result of bioprospecting in the forests of developing countries and the latter countries have received little benefit from this.
3.6i	B	Island biogeography examines the factors that affect the colonisation of islands by species and the extinction of island species. The principles established have been used to examine nature reserves as if they were islands.
3.7i	A	Domestic (feral) cats have been responsible for the demise of many island species and species of marsupials in Australia. Many islands are home to populations of feral cats that have descended from cats abandoned by whalers, sealers and other temporary populations. They have been particularly successful in places where the local fauna is not adapted to the presence of predators.
3.8i	C	*K*-selected species include elephants, dinosaurs and large primates. Their populations fluctuate at or near the carrying capacity (*K*) of the environment.
3.9i	B	Peace parks occur in transboundary areas, e.g. the Great Limpopo Transfrontier Conservation Area links national parks in Mozambique, South Africa and Zimbabwe.

3.10i	D	Communities are more likely to value their wildlife and protect it from poachers if it has a commercial value to them by, for example, collecting hunting fees from foreign hunters who pay for hunting licences, accommodation, etc.
3.11i	B	The Royal Society for the Protection of Birds compiles these lists. The IUCN is the International Union for the Conservation of Nature; WWF is the World Wide Fund for Nature; FFI is Fauna and Flora International.
3.12i	B	Critically Endangered (CR), Endangered (EN) and Vulnerable (VU) are the categories of species at greatest risk in this particular order.
3.13i	A	The hotspot approach to conservation focuses attention on a small number of areas of the world and this has the effect of attracting funds to these areas while depriving other lesser-known areas (and species) of financial support. This could have the effect of sacrificing large areas of the planet and protecting a very small percentage of the total area.
3.14i	D	Release to the wild may only occur in a very small number of cases at the moment and is not a stated goal of many projects.
3.15i	C	Inbreeding coefficients allow managers to assess the relatedness of a pair of individuals so that mating between close relatives may be avoided thereby reducing the opportunity for inbreeding.
3.16i	B	Rhino horn is not an aphrodisiac but is sold as such in some countries.
3.17i	A	The sequence from most at risk to least at risk is: Extinct (EX), Extinct in the Wild (EW), Critically Endangered (CE), Endangered (EN), Vulnerable (VU), Near Threatened (NT), Least Concern (LC), Data Deficient (DD).
3.18i	C	Approximately three-quarters of all species that have become extinct since 1600 were island species as they are particularly susceptible to disease, have small geographical ranges, are vulnerable to introduced predators, and so on.
3.19i	A	The Sunda pangolin is highly prized in some cultures for its meat and 'medicinal properties'.
3.20i	D	Bees are used in Africa and Asia as a tool in managing human–elephant conflict.
3.1a	D	The minimal viable population can be estimated using population projections calculated by computer using population viability analyses.
3.2a	C	An ESU is an evolutionarily significant unit and is considered distinct for conservation purposes.
3.3a	C	Captive breeding programmes for a species generally include animals from different zoos that are bred together to increase the overall population size so they are effectively a single population known as a metapopulation.

3.4a	B	This is the rescue effect. The distracters are fictitious terms in this context.
3.5a	D	Jump dispersal may occur due to human action. For example, a species with a low inherent ability to disperse may be spread around the oceans on the bottom of ships or by air transport. The distracters are fictitious.
3.6a	B	A is better than B because the areas are closer together allowing more movements of species between them and less chance of leaving a protected area and arriving somewhere that is unprotected. D is better than C because in D the areas are grouped together and in C the areas at both ends of the row are a long way from each other.
3.7a	C	Birds on islands by definition have a restricted range and those in mountainous areas in the tropics are similarly isolated.
3.8a	A	Catastrophic mass extinctions of birds on Pacific Islands followed the arrival of the first humans.
3.9a	C	Contingent means existing only under certain circumstances.
3.10a	A	Conservationists need to strike a balance between conserving as much biodiversity as possible within a species and creating so many ESUs that none has a viable population and hybrids between proposed candidate ESUs become useless for future breeding.
3.11a	C	Conservation International requires a hotspot to possess at least 1,500 endemic vascular plants and have 30% or less of its original vegetation.
3.12a	A	Point B is the species richness achieved when the immigration rate and the loss rate from extinction are in balance.
3.13a	D	CRISPR stands for clustered regularly interspaced short palindromic repeats.
3.14a	D	The distracters are fictitious.
3.15a	C	A genetic bottleneck reduces the size of the gene pool and eliminates some genotypes resulting in some genes being removed from the population.
3.16a	B	These structures are bird diverters and are designed to prevent large birds from colliding with power lines. This power line is located near a nature reserve owned by the Wildfowl and Wetland Trust (WWT) in England that is visited by large numbers of migratory geese and swans.
3.17a	C	A cloned Przewalski's horse foal was born to a surrogate in 2020. It was conceived using sperm that had been cryopreserved since 1980 by San Diego Zoo Global.
3.18a	A	This is a small isolated population and it is likely that the animals are inbreeding, leading to the appearance of abnormalities.

3.19a	A	The Clovis culture spread via a land bridge from Asia to North America and then South America probably bringing new hunting techniques. Large numbers of endemic American taxa disappeared as a result.
3.20a	D	These were historically important events and the British Prime Minister at the time, Lord Salisbury, addressed the government about the problem. Several films have been made about the maneaters based on the book by the hunter J. H. Patterson.

Chapter 4 Restoration Biology and Habitat Management

4.1f	C	The first animals used for breeding are the founders.
4.2f	B	Grey wolves have been dispersing naturally across Europe for many years. The term 'dispersion' is incorrect here as it refers to the geographical spacing of individuals.
4.3f	A	Beavers alter water courses by building dams. The resultant change in the landscape changes the biological community. The distracters are fictitious in this context.
4.4f	C	Rabbit fences exist across parts of Australia to keep rabbits away from pastoral areas.
4.5f	C	Submerged metal frames have been widely used to act as a 'skeleton' upon which corals can grow.
4.6f	D	Natural capital is correct and refers to all of the assets the Earth provides.
4.7f	C	Afforestation refers to planting trees in an area that has not previously been forested.
4.8f	A	Hydroseeding is a means of rapidly seeding an area by spraying it with a slurry of seed and mulch. Broadcast seeding involves scattering seed by hand or using a machine. Seed drilling involves using a machine to insert seeds into the soil. Dibbling is the act of making a small hole with a tool called a dibble and placing a seed or seeds within the hole.
4.9f	D	This is a translocation because animals have been moved from one place to another. An 'introduction' would imply that no elephants previously lived in Tsavo. A 'reintroduction' would suggest that the elephants in Tsavo had become extinct and were being put back.
4.10f	C	C is correct. The distracters are simply the same terms rearranged.
4.11f	B	A hard release would not include a period of adjustment in a holding pen. The other distracters are fictitious.

4.12f	A	An ecosystem service is a function provided by an ecosystem that benefits humans, such as flood protection. Recognition of ecosystem services helps to provide ecosystems with an economic value.
4.13f	A	The project was initiated by Phoenix Zoo in Arizona.
4.14f	D	Shelterbelts exist at the edge of many deserts, for example at the southern edge of the Sahara in northern Nigeria to impede the southward movement of sand.
4.15f	A	Remediation refers to the reversal of environmental damage.
4.16f	B	Mitigation refers to the act of reducing the severity of something.
4.17f	D	The original forest in an area is primary forest. If this is removed and then replaced the new forest is called secondary forest.
4.18f	C	Undisturbed fish in the MPA breed, increase in number and disperse into the adjacent fishing grounds.
4.19f	A	Mycorrhizal fungi colonise the roots of plants and live in a symbiotic association with them, helping to supply nutrients.
4.20f	D	Biosorption is the ability of organisms to accumulate heavy metals and remove them from water (e.g. yeasts, seaweeds and fungi).
4.1i	C	This approach improves the chances that released animals will survive. The distracters are fictitious in this context.
4.2i	A	Mowing too early has the effect of damaging flowering plants at an early stage in their growth and reducing invertebrate diversity.
4.3i	B	Wetland carbon farms have the potential to store large quantities of carbon while also protecting a habitat type that has suffered a spectacular global loss of area.
4.4i	A	Ecoduct is a collective term for structures that allow wildlife to cross barriers to their movements.
4.5i	C	Heather is burned to provide open areas where new shoots can emerge, providing food for the grouse. It also kills trees and grasses.
4.6i	B	Reforestation is correct. Afforestation refers to the planting of trees in an area that has not been forested. Forestation may refer to reforestation or afforestation so is less specific.
4.7i	B	The process is only considered to be a translocation if it replaces a pre-existing habitat of the same type.
4.8i	C	This is population supplementation because the released individuals are joining an existing population.
4.9i	A	In some parts of the world (e.g. the east coast of England) the coastline is receding due to the action of the sea. This is expensive and, in some places, impossible to stop, so affected communities must be relocated.

4.10i	C	Sand dunes form when an onshore wind blows across a wide expanse of sand and sand particles collect around objects on the beach. Dead trees and low wooden fences placed on the sand encourage sand dune formation.
4.11i	B	Coniferous trees are planted in the watersheds of reservoirs primarily to intercept rainfall so that it drips off the trees over an extended period of time. If reservoirs fill too quickly water may be 'wasted' by allowing it to pass into rivers via overflow devices.
4.12i	B	Amenity grasslands are generally mown to a short length and treated with chemicals. They may contain drainage to prevent waterlogging and biodiversity is low.
4.13i	C	The process was called 'Pleistocene rewilding' as this was the time when similar animals roamed this area. These animals would be ecological analogues because they would perform a similar ecological role to related taxa that would exist in the United States if their ancestors had not become extinct.
4.14i	D	D is correct. The distracters are the same processes in different orders.
4.15i	C	The temperature during the day in the forest at A would be higher than that at B because A is more exposed to the sun.
4.16i	D	Many small mammal populations are seriously damaged by fires.
4.17i	B	They cannot be dried or frozen so are unsuitable for cryopreservation in seed banks.
4.18i	A	When a habitat is restored it is important for ecologists to base their management actions on a reference ecosystem that is like the system they are hoping to recreate.
4.19i	D	The longhorn is a rare breed of cattle and is used to manage grassland areas, for example in nature reserves inhabited or visited by large flocks of waterfowl.
4.20i	B	This waste is referred to as tailings. Overburden is the surface material removed to expose the ore. The other distracters are irrelevant terms used in mining.
4.1a	C	The term 'Judas animal' refers to Judas Iscariot and the term 'Judas' is used to refer to anyone who has acted deceitfully.
4.2a	B	The debate is about whether one large protected area is better for conserving wildlife than several smaller areas.
4.3a	C	Ecological analogues have a similar ecological function and may be used in ecosystem restoration to replace services previously provided by lost species.
4.4a	B	Vernalisation refers to a process that occurs during seed germination. The other distracters are fictitious in this context.

4.5a	D	Removing eggs may result in birds producing a second clutch thereby potentially increasing reproduction rates. Transferring eggs to an incubator prevents losses from predation (e.g. by mink, foxes, etc.) and increases hatching rates.
4.6a	A	The nuns have assisted with the conservation of these salamanders by creating a breeding programme together with Chester Zoo in the UK, the Mexican Government fisheries centre, and the Universidad Michoacana de San Nicolás de Hidalgo, Mexico.
4.7a	B	Metalophytes are varieties of plants that have evolved resistance to heavy metals in the soil that would otherwise be toxic to them.
4.8a	A	Water level is lowered during the breeding season to expose a larger area of land for breeding around the shoreline of the mere and its associated islands.
4.9a	B	The process begins with a founding phase where a small number of animals are used for the initial breeding (the founders). The population is then allowed to grow (growth phase) and eventually reaches the capacity phase when the population reaches its target size.
4.10a	D	This is a top-down cascade of effects because it began with the reintroduction of a top predator.
4.11a	B	This is assisted regeneration because in this scenario ecologists help the restoration process by fostering natural regeneration with limited human intervention.
4.12a	D	All stakeholders should be consulted, i.e. anyone that could benefit or be disadvantaged as a result of the introduction, especially if the species concerned has the potential to affect farming, forestry, fishing or other aspects of the local ecology and economy.
4.13a	C	Individual sites exist in a larger landscape so 'landscape flows' is correct.
4.14a	A	Genetic drift is the process whereby genes are lost from a population due to chance events. If the population consists of just 10 individuals and only one of these possesses gene A, a storm that kills 50% of the population has a 50% chance of completely removing the gene from the population.
4.15a	A	Outbreeding depression may occur if distantly related individuals interbreed.
4.16a	D	The species must be sympatric (not allopatric), i.e. occur in the same geographical area, or they will not be able to meet and mate. They must not be mechanically isolated, i.e. they must have compatible genitalia. They must be closely related so that they can produce a viable embryo and they must be able to exhibit appropriate courtship and mating behaviour.

4.17a	B	A phosphate sorbent is a chemical that could be used in a eutrophic lake to absorb phosphate.
4.18a	A	Some topsoil is stored wet and some is stored dry. The distracters are fictitious.
4.19a	B	Some trees have mycorrhiza (symbiotic fungi) associated with their roots that help them obtain nutrients. Stockpiling topsoil reduces the extent to which these fungi are available.
4.20a	B	This is a method of captive breeding and release whereby animals are kept in large enclosures and allowed to rear their young before eventually releasing the young from the enclosures to the wild.

Chapter 5 Agriculture, Forestry and Fisheries Management

5.1f	B	Dolphins and other cetaceans are regularly, but unintentionally, caught and killed as bycatch in fishing nets.
5.2f	D	Mariculture is the most accurate term for this; however, it is a type of aquaculture.
5.3f	C	In many developing countries farmers produce crops that are largely sold to developed countries (e.g. coffee, tea, palm oil) rather than food to feed themselves. These crops are called 'cash crops'.
5.4f	C	Hydroponic systems grow plants in a nutrient solution in glasshouses without the use of soil.
5.5f	A	Silviculture is the care and cultivation of woodlands. Coppicing involves cutting down trees to ground level to stimulate growth; pollarding is a woodland management method used to encourage the lateral growth of branches.
5.6f	A	As wind blows across the leaves of a plant it causes water to be lost by the process of transpiration.
5.7f	D	Crushed limestone is calcium carbonate.
5.8f	B	Some power stations feed warm water to large fish tanks. Keeping the water warm encourages their growth.
5.9f	D	In some agricultural systems animals are moved typically to lowlands in winter and highlands in summer, e.g. in the mountains of Norway.
5.10f	B	A fish ladder allows fish to pass obstructions in a river and typically consists of a series of stepped pools through which the water flows.
5.11f	A	Aquaculture is a general term for farming aquatic organisms; mariculture is the farming of marine organisms.
5.12f	D	The private ownership of land is an important way of protecting it.

5.13f	C	Softwoods (e.g. gymnosperm trees) grow faster, not slower, than hardwoods (angiosperm trees).
5.14f	A	Raymond Dasmann was an important conservationist who pioneered the concept of game ranching. The distracters are other well-known ecologists.
5.15f	C	The hidden text says 'Rainforest Alliance'. The distracters are fictitious.
5.16f	B	Vernalisation initiates flowering in some species. The distracters have no meaning in this context.
5.17f	B	Agroforestry is a more sustainable method of farming than simply growing a single crop.
5.18f	A	Ash dieback is a chronic fungal disease of ash trees in Europe.
5.19f	B	Beech trees are angiosperm trees so are hardwoods. The distracters are coniferous so are softwoods.
5.20f	D	Domestication is a genetic process whereby humans select, and breed from, organisms that have desirable properties such as high milk yield, tameness, etc.
5.1i	D	Although whales occur in the North Sea it does not possess a whale sanctuary.
5.2i	C	Rotational grazing prevents the same areas from being grazed continuously.
5.3i	D	Intercropping is an ecologically beneficial method of farming compared with producing single crops.
5.4i	A	A seine net has a fine mesh and is suspended vertically in the water with weights holding it to the bottom. Such nets are banned in many parts of the world because of the damage they do to fisheries.
5.5i	B	After point B on the graph the yield (catch) decreases if fishing effort is increased.
5.6i	D	It is important for farmers to know the digestibility of grass and how this changes as the plant ages because this will determine the nutritional value of the grass to livestock.
5.7i	B	When soil is saturated with water the oxygen content falls as the pore spaces are filled with water.
5.8i	A	Whalers were allowed to catch a certain number of blue whale units under a system of quotas. One blue whale = two fin whales = six sei whales.
5.9i	B	The Trust maintains a 'watchlist'. A 'Red List' of species is produced by the IUCN.
5.10i	D	Magnesium is an important component of the structure of chlorophyll so if it is deficient in the soil chlorosis will result. It can also be caused by poor soil drainage and exposure to sulphur dioxide pollution in the air.

5.11i	C	Legumes (e.g. beans, peas and clover) have roots that bear nodules containing nitrogen-fixing bacteria so are grown in rotation to replace the nitrogen removed by other crops.
5.12i	A	A is correct. The distracters are fictitious in this context.
5.13i	C	This is a rare breed of goat.
5.14i	A	Rainfall can add nitrogen. The distracters all remove it or make it unavailable.
5.15i	C	Young fish would escape and continue to grow so eventually more large (older) fish would be caught.
5.16i	B	Strip grazing prevents overgrazing. It may be possible to allow cattle to feed on a field all year if they are only allowed access to narrow strips of land at any one time, thereby allowing other areas to recover.
5.17i	D	All of these options may be a consequence of waterlogging.
5.18i	B	If several species of wild herbivores graze together they are likely to occupy different feeding niches so do not compete for the same plant species.
5.19i	B	When fish are being grown to marketable size the process is called 'growing on'. This is effectively the equivalent of the term 'finishing' in livestock farming.
5.20i	A	Sea lice are copepods. The distracters are other groups of cructaceans.
5.1a	A	A reduction in spray pressure should reduce spray drift. The distracters would all increase spray drift.
5.2a	B	B is correct. The distracters are the same options rearranged.
5.3a	C	Fires burn more rapidly up a slope (not down) because the wind moves more rapidly up slopes (not down).
5.4a	C	$5/0.22 = 22.7$, approximately 23.
5.5a	D	All of these fisheries have collapsed.
5.6a	D	All four options are possible.
5.7a	D	The shelterwood method involves a small number of cuts over a long period so that young trees are always protected by older trees.
5.8a	A	Intraspecific competition (within a species) is at play at the highest density not interspecific competition (between species).
5.9a	B	Seed ratio = grains seeded: grains harvested.
5.10a	B	Methyltestosterone is a sex hormone and is used to convert females to males. Males tend to have higher growth rates and carcass yield than females.
5.11a	C	This was a long-term study conducted at Rothamstead. The distracters are the names of real research stations but they were not involved in the study.

5.12a	D	The term for this phenomenon is 'critical *depensation*' not 'critical *dispensation*'.
5.13a	D	Although this announcement was made this will make little difference to the total methane output from cattle.
5.14a	B	Leaf area duration is a measure of the retention of green leaves with time: $m^2/m^2/days$.
5.15a	D	In Glasshouse 1 the population fluctuates because chemical control is being used. In Glasshouse 2 the population is being maintained at a relatively constant low level because it is being controlled by the predators.
5.16a	D	Nitrification is an aerobic process so requires oxygen.
5.17a	A	The prefix 'pharm' in the term 'pharming' relates to pharmaceutical. Pharming involves the creation of GM organisms.
5.18a	B	Large logs on the forest floor play little part in sustaining forest fires. Leaves help to retain moisture on the forest floor.
5.19a	A	Hypophysation is used to induce spawning in fish farming and involves the injection of gonadotropins.
5.20a	D	Drones are now being widely used in ecological monitoring and agriculture.

Chapter 6 Pest, Weed and Disease Management

6.1f	B	This disease has had an enormous effect on amphibian populations worldwide and has led to increased efforts to conserve many species *ex-situ*.
6.2f	A	Sticky traps catch insects when they come into contact with an adhesive.
6.3f	B	Anthelmintic drugs are used to treat infections with internal parasitic worms and cause them to be expelled from the body.
6.4f	D	2,4-D is a systemic herbicide called 2,4-Dichlorophenooxyacetic acid.
6.5f	A	'Zoo' refers to animal; 'nosis' refers to disease.
6.6f	D	Some plants produce chemicals known as 'antifeedants' which deter insects or other animals from feeding on them.
6.7f	A	CWD affects deer, moose, caribou and their relatives.
6.8f	A	Ebola is a rare deadly disease of primates, including humans.
6.9f	C	The distracters are fictitious in this context.
6.10f	A	Dutch elm disease is caused by a fungus and spread by the elm bark beetle.

6.11f	C	*Phytophthora infestans* is a fungus that causes late blight or potato blight.
6.12f	B	Domestic and feral dogs are responsible for most rabies deaths.
6.13f	B	In this disease nematode worms block the lymph vessels, causing an accumulation of lymph and swelling of the legs.
6.14f	A	The distracters are fictitious in this context
6.15f	D	Wet markets are common in Southeast Asia, especially China, where humans come into very close contact with a wide range of living and dead animals.
6.16f	B	This disease primarily affects cloven-hoofed animals but can also affect elephants.
6.17f	B	TMV is tobacco mosaic virus.
6.18f	B	The individual plants are essentially genetically identical so if one plant is susceptible to a disease all of the others in the field will be too.
6.19f	C	Rabbits were introduced into Australia in 1859. Their population increased rapidly and they damaged grazing land needed for livestock. Mxyomatosis is a viral disease that was introduced in 1950 to control the rabbit numbers.
6.20f	D	The structures are galls on an oak tree caused by a parasitic wasp.
6.1i	C	Insects from the genus *Glossina* transmit protozoans that cause sleeping sickness (*Trypanosoma*). *Anopheles* is a genus of mosquito that transmits malaria caused by the protozoan *Plasmodium*.
6.2i	B	Queleas are small songbirds that feed on grain crops in very large flocks.
6.3i	A	A chemosterilant is a chemical that causes reproductive sterility.
6.4i	C	Birds are the main vector of dengue virus. Mosquitoes become infected when they feed on infected birds.
6.5i	B	Hydrogen cyanide is a first generation pesticide.
6.6i	A	2,4-D is 2,4-Dichlorophenoxyacetic acid, a synthetic auxin.
6.7i	D	The second crop acts as a trap for the pests by attracting them away from the primary crop.
6.8i	B	This species suffers from a contagious cancer – Tasmanian devil facial tumour disease. Wildlife managers have erected fences in parts of Tasmania to try to prevent its spread.
6.9i	A	Wheat, barley and oats shed seeds that may germinate and form volunteer populations in other crops.
6.10i	D	European badgers do not occur in Australia. The other species all have feral populations.

6.11i	B	The original infestation (red scale) is a primary outbreak. The increase in numbers of the second pest (white wax scale) caused by human intervention is a secondary outbreak.
6.12i	D	Transgenic because a gene has been taken from one species and inserted into another. A chimera is an organism that has two or more distinct genomes. A clone is an organism that is genetically identical to another. Pangenic plant is a fictitious term.
6.13i	C	The cactus was introduced into Australia in 1839 and spread rapidly. The moth was introduced into Australia from Argentina in 1925.
6.14i	C	The term 'close season' is used in ecological management for a period when a particular activity is not permitted, e.g. hunting and fishing. Sometimes also called a 'closed season'.
6.15i	B	Sterile insect technique (SIT) is an environmentally-friendly method of controlling insects that involves releasing sterile males into a pest population where they mate with wild females but no offspring result.
6.16i	D	The definitive host is the species where the adult worm lives. In this case it is the bush dog and the domestic dog.
6.17i	D	Newcastle disease affects birds and is caused by a virus.
6.18i	A	The source is not absolutely certain but is believed to be bats.
6.19i	C	Rinderpest is an important viral disease.
6.20i	A	The water hyacinth is a floating flowering plant that occurs widely in Lake Victoria in East Africa. It is native to South America.
6.1a	D	HIV is correct (human immunodeficiency disease). SARS is severe acute respiratory syndrome. MERS is Middle East respiratory syndrome. EID is an acronym for emerging infectious diseases.
6.2a	C	Cross resistance is a tolerance to one toxic substance that results from exposure to a substance with a similar action and occurs widely in pesticides, herbicides and antibiotics.
6.3a	B	LD_{50} is the dose that kills 50% of the organisms, in this case $47mg/m^2$.
6.4a	C	When a pest reinvades an area that has been sprayed due to its natural enemies having been killed off this phenomenon is known as resurgence (a revival following a period of little occurrence).
6.5a	A	The chance appearance of a natural enemy of a pest that subsequently controls that pest is known as fortuitous biological control. Fortuitous means happening by chance rather than intention. The distracters are fictitious in this context.
6.6a	D	The sequence must begin with a host. Hyperparasitoids parasitise parasitoids. D is correct. It is not necessary to know the identity of A, B or C, but A is a moth, B and C are both hymenopterans.
6.7a	D	The Greek root *all-* means 'other' and *path-* relates to disease.

6.8a	B	Ducks and gulls are responsible for the majority of bird strikes. They gather in large flocks. Gulls frequently rest on grass runways and on grassed areas around large airports.
6.9a	D	Species of all of these organisms have been used to control weeds.
6.10a	B	This insect originated in Asia and has killed millions of ash trees (*Fraxinus*) in Canada. The distracters are other wood-boring beetles.
6.11a	A	DDT, Lindane and Aldrin are organochlorines; Malathion is an organophosphate. Lists B, C and D only contain organochlorines.
6.12a	A	Using unrelated classes of insecticides in rotation makes it more difficult for resistance to evolve in insects.
6.13a	C	Adult parasites infect the liver.
6.14a	D	Citrus oils are used to make organic insecticides and acaricides (e.g. extracts from orange peel, lemon oil).
6.15a	D	Phytosanitary certificates are issued when plants or plant products are to be exported to another country and certify that the material is free from harmful pests and diseases. The purpose is to prevent (or at least reduce) the spread of pests and diseases around the world.
6.16a	A	In the EU the European Food Safety Authority is the competent authority. The distracters are other EU organisations that are not involved in the safety of plant protection products.
6.17a	B	APHIS is the Animal and Plant Health Inspection Service. APHID is fictitious. ADAS is the former Agricultural Development Advisory Service (now a UK agricultural and environmental consultancy). USDA is the US Department of Agriculture. APHIS is an agency of USDA.
6.18a	A	These are notifiable diseases and pests. The distracters are fictitious in this context.
6.19a	C	These are all types of mechanical weed control. Mechanised control implies that a machine is used. Machines are sometimes, but not always, used in mechanical control.
6.20a	D	If black plastic sheets are placed over the plants they will deprive them of light and raise the temperature thereby killing them. Clear plastic may also be used.

Chapter 7 Urban Ecology and Waste Management

7.1f	A	A is correct. The distracters are fictitious.
7.2f	D	A brownfield site is a disused site of previously developed land that may contain some historical contamination related to its previous use.
7.3f	D	A relic (or relict) community is a surviving remnant of a biological community that has largely disappeared from the area.

7.4f	C	City centres generally contain large and tall buildings containing ledges and high roof tops, resembling cliffs.
7.5f	B	The concrete and other building materials, roads, bridges, etc. in a city centre absorb and retain heat more than surrounding rural areas.
7.6f	A	A synanthrope lives alongside humans and their habitats, e.g. house sparrows, rats, pigeons, cockroaches, raccoons.
7.7f	A	Large flocks of starlings were once common in European cities and could be seen (and heard) roosting on window ledges and roofs in places like Manchester, England. Their aerial displays are called murmurations.
7.8f	B	Anthropogenic means caused by humans: on vehicle tyres, boots and shoes and in topsoil and rubble moved by people in the course of construction, etc.
7.9f	C	Green roofs slow down runoff as rain is initially intercepted by plants and absorbed by soil.
7.10f	B	Mosquitoes breed in water.
7.11f	A	Trees absorb solar energy and prevent it from reaching the soil surface. The albedo of an area cleared of soil will increase because energy will be reflected from the ground.
7.12f	C	Dispersal is the ability to spread over a wide area. Seeds carried by birds and the wind have germinated on various parts of this building.
7.13f	B	Salt-tolerant plants have colonised these areas, e.g. grasses of the genus *Puccinellia*.
7.14f	A	The UK has a large number of private gardens because of the type of housing stock. Many other countries in Europe have a much higher proportion of apartments within large buildings with no gardens.
7.15f	D	A bioswale is essentially a vegetated drainage ditch that removes sediment and pollution from rainwater.
7.16f	C	In 1966 a colliery spoil heap in Aberfan collapsed and buried a school, killing a large number of people, mostly children.
7.17f	B	Garden cities first appeared in England in the 1900s. Some were built by employers for their staff.
7.18f	D	Breaks in the sewerage pipes allow groundwater to enter and increase the volume of the sewage. This process is called infiltration.
7.19f	B	Bird spikes are groups of fine metal rods that prevent birds from landing on the tops of street lights, window ledges and other urban structures.
7.20f	D	Rodents are the most important vectors of sylvatic plague.
7.1i	B	Urban avoiders do not completely shun urban areas but are more successful in natural sites.

7.2i	A	Foxes tend to have smaller ranges in urban areas than in rural areas as there are more food sources to exploit, e.g. pet animals, waste food, etc. Urban foxes have been extensively studied, e.g. in Oxford and Bristol in England.
7.3i	D	The activated sludge process was first used at Davyhulme Sewage Treatment Works in Manchester, England in 1914.
7.4i	D	Ruderal is correct. Benthic species live at the bottom of bodies of water. Lotic species live in rapidly moving water. Littoral species live on shores.
7.5i	B	Buzzards are generally found in open farmland areas and would find it difficult to hunt in urban areas.
7.6i	C	Introducing plants of different heights increases structural complexity and increases the number of available niches.
7.7i	D	It is essential to prevent the cocktail of chemicals in landfill waste from leaching out into the surrounding soil, aquifers and watercourses.
7.8i	C	An oxidation ditch is a relatively simple small sewage treatment plant consisting essentially of a concrete tank containing sewage that is aerated by a large brush-like structure rotating around its horizontal axis.
7.9i	D	All of these effects occur in some species.
7.10i	D	All of the processes listed have a negative effect on ecosystems (provide a disservice) as opposed to, for example, a sand dune system that provides a flood defence (i.e. an ecosystem service).
7.11i	A	Bacteria in an anaerobic digester break down sludge and produce methane in the process. This is used to generate electricity in some wastewater plants.
7.12i	C	Modern towns and cities have separate sewerage systems: rainwater runoff from roads and roofs flows untreated into the local river, domestic and industrial waste enters a separate system that carries it to the local wastewater treatment works. This prevents the plant from being overloaded with rainwater that does not need to be treated.
7.13i	B	Allotment gardens are small parcels of land that are allocated to local residents and used to grow vegetables, fruits and other plant produce. The variety of species grown and the unkempt nature of these sites encourages colonisation by a wide range of native animals and plants.
7.14i	D	Sewage sludge should not be sprayed on land immediately before it is to be used by grazing animals nor sprayed on fruit. Sludge from industrial areas must be carefully monitored for toxic metals and the metal levels in soils treated with sludge must also be monitored.
7.15i	A	In the 1970s petrol contained lead. This was spilled on roads, at petrol stations and garages, etc., and then washed into the sewerage system during storms.

7.16i	D	PTEs stands for potentially toxic elements (e.g. cadmium, mercury, lead).
7.17i	C	This is known as the first flush. The distracters are fictitious in this context.
7.18i	A	'Urban metabolism' is correct. The distracters refer to other metabolic processes and have no meaning in this context.
7.19i	B	The traits referred to are aspects of the animal's personality.
7.20i	C	Exfiltration is effectively the opposite of infiltration (the inward movement of water). The distracters are fictitious in this context.
7.1a	D	The Chicago School of Sociology refers to the work of staff and graduate students working at the University of Chicago, including work on urban ecology and applied research within the urban environment.
7.2a	B	Such soils contain very little organic matter so are deficient in nitrogen.
7.3a	C	Many lichens are intolerant of air pollution, particularly sulphur dioxide, so lichen diversity increases with distance from the city centre.
7.4a	D	D is correct. The other terms are fictitious.
7.5a	B	The activated sludge process uses aeration tanks so A and C can be eliminated. Option D has the correct components in the wrong sequence so B is correct.
7.6a	A	The screen removes large materials from the raw sewage before it reaches the primary settlement tanks. This prevents blockage and damage. Towels are accidentally washed down sluices in hospitals and large tree branches also find their way into some sewers (along with glasses and false teeth!).
7.7a	C	This is a photograph of a London plane. The distracters are all tree species found in Britain.
7.8a	B	This graph shows a positive correlation, i.e. as distance increases the number of bird species increases.
7.9a	B	This is a storm water retention tank used to intercept high sewage flows during and immediately after storms to prevent the material reaching the primary settlement tanks. A very high flow to the primary tanks would reduce their retention time and consequently their efficiency at removing suspended solids. The storm water is pumped to the primary tanks for treatment after the storm has passed. Normally storm tanks are kept empty until they are needed.
7.10a	C	C is correct. The other options have 'higher' and 'lower' arranged in a sequence that has no specific meaning.
7.11a	B	B is correct. The distracters produce a table that is inaccurate.
7.12a	D	A circular urban metabolism is more environmentally-friendly than a linear urban metabolism (see Fig. 7.2) because there is less waste of energy and materials and lower pollution emissions.

7.13a	C	The temperature is high in the built-up areas and lower in the rural areas and the park. The hottest area is the central business district as it contains a high density of large buildings and roads. The buildings and roads absorb and retain heat, whereas a higher proportion of solar energy is reflected from rural and parkland landscapes. The distracters have no meaning and are simply the correct temperature profile rotated through its vertical and horizontal axes.
7.14a	B	30,000 litres/sec x 60 x 60 x24 = 2,592,000,000 litres/day. This value divided by 1000 = 2,592,000 m³/day, or approximately 2.6 million m³/day.
7.15a	C	DWF = (catchment population x per capita domestic flow) + dry weather infiltration + trade effluent flow. Storm water runoff and ambient temperature are irrelevant.
7.16a	A	The map showed that chloride concentrations in ground and surface waters increased as one nears the Atlantic coast. Deviations from 'normal' levels indicated sewage contamination. Richards produced the world's first water purity tables.
7.17a	B	The social organisation of feral cats is similar to that of lions. Females and their young live in social groups, sometimes with resident adult males, with other nomadic adult males moving between the groups.
7.18a	B	The largest bird species tend to dominate feeders.
7.19a	A	Such projects are often called citizen science projects and are capable of generating large quantities of data.
7.20a	D	At a fine scale the environment was heterogeneous and this allowed many species to survive across a wide range of habitats.

Chapter 8 Global Environmental Change and Biodiversity Loss

8.1f	B	Carbon dioxide is the most important greenhouse gas.
8.2f	D	Al Gore featured in An Inconvenient Truth (released in 2006). The distracters are also former US Vice-Presidents.
8.3f	B	Carbon footprint is correct. The distracters are fictitious in this context.
8.4f	B	Sulphur dioxide is produced during the combustion of coal and produces acid rain when released into the atmosphere. A sulphur dioxide scrubber is a device that removes SO_2 by causing it to react with a mixture of limestone and water.
8.5f	A	A is correct. The distracters are other organisations concerned with the environment and conservation.

8.6f	D	This term usually refers to the cancelling out of carbon emissions by absorbing an equivalent amount by, for example, planting forests.
8.7f	D	Most of the atmosphere (78%) is nitrogen. It is not a greenhouse gas.
8.8f	C	Tundra contains very little plant material so has little potential to absorb carbon dioxide.
8.9f	A	Millions of animals, especially reptiles, were killed by extensive bushfires in the 2019-2020 bushfire season.
8.10f	A	The purpose is to reduce greenhouse gas emissions. Methane does not have an odour.
8.11f	D	All of these sectors produce carbon dioxide, methane and other greenhouse gases.
8.12f	B	Amphibians were the most threatened group by this measure.
8.13f	D	Specialist feeders cannot easily exploit new habitats and those that produce only a single generation per year cannot evolve and adapt as quick as those species that produce more than one generation per year.
8.14f	A	In the Bible, Lazarus rose from the dead. The distracters are fictitious but with biblical references.
8.15f	A	A is correct. The distracters are fictitious.
8.16f	D	Meltwater adds water to the oceans, seawater expands as the climate warms and tectonic activity changes the shape of ocean basins while volcanic activity releases greenhouse gases.
8.17f	A	The LPI measures trends in vertebrates only.
8.18f	C	C is correct. The distracters are other well-known Swedish scientists.
8.19f	C	C is correct. The distracters list high emitters but in the wrong order.
8.20f	B	An increase in volcanic activity appears to be associated with all but one of the mass extinctions.
8.1i	B	Dendrochronology uses the thickness (growth) of tree rings to deduce changes in the climate with time.
8.2i	B	In B emissions are 21+56=77. Absorption =77. In the distracters emissions and absorption are not in balance.
8.3i	C	Northern and central Europe are warming so crop-growing conditions should improve here.
8.4i	B	Buffering is correct. The distracters are words with a similar meaning.
8.5i	B	Sequester means 'to temporarily take possession of' or 'seize'. Trees temporarily remove carbon from the atmosphere.
8.6i	A	British Antarctic Survey scientists discovered the 'hole' in the ozone layer using a Dobson Ozone Spectrophotometer and its existence was later confirmed by NASA satellites.

8.7i	B	Organisms that are *r*-selected have a short generation time, expand their populations quickly and exploit new environments. Short generation times allow faster evolution.
8.8i	D	Polar bears are spending more time on land, looking for new food sources and their movements are becoming more difficult to predict thereby increasing conflict with humans.
8.9i	A	Industries that reduce their emissions below the level permitted can trade their permits with those that need to increase their emissions. This system is used within the European Union in an attempt to slow climate change.
8.10i	D	Scientific evidence shows that the predicted increase in sea level rise in the 1970s did not occur and this can only be explained by the increased amount of water being prevented from reaching the sea by new dams and reservoirs built around the world at that time.
8.11i	D	D is correct. The distracters are fictitious.
8.12i	B	El Niño causes disruption to global climate leading to storms and droughts.
8.13i	A	Snow arrives later leaving roots exposed, the winter is shorter resulting in a shorter hibernation, and range extends northward because temperatures are rising.
8.14i	A	A 'tipping point' refers to the situation whereby the cumulative effect of a series of small changes becomes significant enough to cause a much larger change. Specifically it refers to the point in time at which the effect cannot be stopped.
8.15i	B	Biocapacity deficit is correct. The distracters are fictitious in this context.
8.16i	C	Stochasticity is the quality of lacking any predictable order. Small populations are vulnerable to random events and prone to genetic drift.
8.17i	B	Human population increases are associated with increases in mammalian extinction rates.
8.18i	C	There have been five mass extinctions.
8.19i	D	The report estimated that 1,000,000 species were threatened with extinction.
8.20i	B	Global sea level is measured using satellites. Local sea level is measured using tide stations where water level is measured with respect to a specific point on the land.
8.1a	A	Some ecologically distinct species are highly threatened (e.g. Asian elephants) while others are successful widespread generalists that thrive in many different environments, e.g. the lesser black-backed gull).
8.2a	C	Carbon dioxide from the atmosphere dissolves in seawater to form carbonic acid.
8.3a	B	$((414-316)/316) \times 100 = 31\%$.

8.4a	B	This relationship allows us to infer past climatic conditions from the plant fossil record.
8.5a	C	230/(2013-1880) = 1.73mm/year.
8.6a	D	D is correct. The distracters are other major observatories that are not involved. Mauna Loa was selected because it was hoped that its location would provide data that are representative of the Earth's atmosphere as a whole.
8.7a	B	As Europe's climate warms the midges that transmit bluetongue have moved northwards.
8.8a	C	'Extinction debt' refers to the future extinctions that are inevitable due to past events: delayed extinction.
8.9a	D	Some industries have been allocated carbon emission permits on the basis of their past emissions. In legal terminology 'grandfathering' refers to an exemption from a new regulation.
8.10a	C	The people of Isle de Jean Charles have been described as the world's first climate refugees following the allocation of a $48 million federal resettlement grant in 2016.
8.11a	B	Permafrost is permanently frozen ground in the Arctic. As it melts methane and carbon dioxide are released into the atmosphere as plant material decomposes.
8.12a	A	The five extinctions are sometimes described as occurring as follows: Ordovician-Silurian, Devonian, Permian-Triassic, Triassic-Jurassic and Cretaceous-Tertiary.
8.13a	D	These are all 'Lazarus species': species which were once thought to be extinct in the wild but have been rediscovered. Rare species are by definition hard to find so it is not surprising that this occurs from time to time, especially in remote areas.
8.14a	B	B is correct. The distracters are other Antarctic glaciers.
8.15a	C	The graph shows that by 2016 the LPI value had dropped to 68% of its value in 1970. It refers only to trends in vertebrate species biodiversity.
8.16a	D	The IPBES was established in 2012 to strengthen the science–policy interface for biodiversity.
8.17a	B	The Aichi Targets were set by Parties to the UN Convention on Biological Diversity to address the underlying causes of biodiversity loss.
8.18a	A	A is correct. The distracters are other remote locations.
8.19a	B	Elephants are predicted to move upwards where the temperature will be lower and the rainfall higher. The supply of food and water should be more reliable at higher altitudes.
8.20a	D	Seawater is flooding land making it difficult to cultivate, beaches are suffering from erosion and the traditional lifestyle of the indigenous people is under threat.

Chapter 9 Environmental and Wildlife Law and Policy

9.1f	D	Many cultures recognise sacred mountains, rivers and forests and may prohibit certain activities within them, such as felling trees.
9.2f	A	A closed (or close) season is a period when hunting is illegal.
9.3f	C	This law regulates trade in animals and plants, whether dead or alive, along with their eggs, seeds and other propagules, and products made from the ivory, skins and other parts of listed species.
9.4f	C	Special Protection Areas are designated under the EU Wild Birds Directive to protect birds and their habitats.
9.5f	B	Natura 2000 is a network of Special Protection Areas (SPAs) designated under the EU Wild Birds Directive and Special Areas of Conservation (SACs) designated under the Habitats Directive.
9.6f	B	The Rasmar Convention is the Convention on Wetlands of International Importance especially as Waterfowl Habitat. It was adopted in Rasmar, Iran in 1971.
9.7f	C	The Parties are the United States, Canada, Norway, Denmark (on behalf of Greenland) and the Former Union of Soviet Socialist Republics (USSR).
9.8f	D	The UNEP was created in 1972 and its headquarters established in Nairobi.
9.9f	C	The moratorium was established from the 1985–86 season onwards.
9.10f	B	Live information about the location of game animals clearly puts a hunter at an advantage over prey animals.
9.11f	A	The precautionary principle requires decisions to be made with caution in the face of uncertainty.
9.12f	B	The United Kingdom passed the Climate Change Act in 2008.
9.13f	A	The Paris Agreement set targets for limiting global warming by reducing greenhouse gas emissions.
9.14f	D	The IPCC was established in 1988 The other options are fictitious.
9.15f	A	CITES has three appendices (I, II and III) and uses a system of export permits from the country of origin and import permits for destination countries.
9.16f	C	The International Commission on Radiological Protection was established in 1928 and provides guidance and recommendations on radiological protection concerning ionising radiation.
9.17f	D	TRAFFIC stands for Trade Records Analysis of Fauna and Flora in Commerce and monitors all aspects of the international trade in animals, plants and their products.

9.18f	D	EUROBATS is an abbreviation for the Agreement on the Conservation of Populations of European Bats 1991.
9.19f	B	The 'polluter pays principle' passes the cost of preventing or cleaning up pollution to the person or organisation that produces it.
9.20f	A	The Earth Summit was held in Rio de Janeiro in Brazil. It resulted in the UN Convention on Biological Diversity 1992.
9.1i	B	In 1989 President Moi destroyed 12 tonnes of elephant ivory in front of the world's press to highlight the plight of elephants in Africa.
9.2i	C	The official forensic laboratory of CITES is located in Oregon in the United States.
9.3i	C	Whales may be taken by a small number of communities who have traditionally practised subsistence whaling and for 'scientific' purposes under the scheme.
9.4i	D	SACs are established in the EU to protect habitats of community concern and European Protected Species of animals and plants, e.g. great crested newts (*Triturus cristatus*).
9.5i	A	CAMPFIRE stands for Communal Areas Management Programme for Indigenous Resources and includes sport hunting and other uses of natural resources. The programme began in the 1980s.
9.6i	B	World Heritage Sites are designated by UNESCO which was founded in 1945.
9.7i	A	The Club of Rome was founded in 1968.
9.8i	D	Species has a broad non-biological definition under CITES that acts as a shorthand for a species, subspecies or geographical population of a species.
9.9i	C	TRAFFIC is an NGO that monitors trade in wildlife. It was created by the IUCN and WWF. MONITOR is a non-profit organisation dedicated to research into trade in lesser-known species.
9.10i	D	Human activity is severely restricted in wildness areas.
9.11i	A	MARPOL stands for the International Convention for the Prevention of Pollution from Ships 1973.
9.12i	A	The strategy was published by the IUCN, UNEP and WWF and considered many of the environmental challenges that we still have today, e.g. destruction of tropical forest, loss of biodiversity, pollution, overfishing.
9.13i	C	This was a series of conflicts between Iceland and the United Kingdom over fishing rights which ended in victory for Iceland in 1976. Iceland established a 200-nautical mile exclusive fishing zone.

9.14i	A	Leg-hold traps are essentially spring-powered steel jaws that trap animals by the legs. Their use has been banned by many countries because this method of trapping is inhumane. Larsen traps are used to live trap wild birds. Longworth traps and Sherman traps are used to live trap small mammals.
9.15i	B	The leaders of 13 tiger range states met to discuss tiger conservation in St Petersburg.
9.16i	C	The convention established a 200-nautical mile exclusive zone as standard around the world.
9.17i	D	Radioactive materials are covered by other laws.
9.18i	C	Joint Aquatic Resources Permit Application (JARPA). This scheme provides whaling permits.
9.19i	B	This Convention is concerned with the spread of deserts.
9.20i	A	The Convention was signed in Bonn, Germany.
9.1a	B	This was a protocol to the Vienna Convention for the Protection of the Ozone Layer 1985 and successfully banned CFCs. These chemicals were responsible for the 'hole' in the ozone layer.
9.2a	A	The case was *Australia v. Japan: New Zealand intervening* (2014).
9.3a	D	Country A sends a shipment to country B with an export licence saying it originated in country A. Country B issues a new export licence (saying B is the country of origin) and sends the shipment to country C. When the shipment reaches country C it has documents saying it originated in country B whereas in fact it originally came from country A.
9.4a	A	A conservation covenant is similar to a contract whereby the landowner agrees to do (or refrain from doing) something in return for money in the form of a tax benefit.
9.5a	D	Some debt-for-nature swaps have involved conservation NGOs paying a debt for developing countries that agree to conserve areas of tropical forest in return.
9.6a	D	The programme had all of these functions.
9.7a	C	The Antarctic Treaty System is correct. The distracters are fictitious in this context.
9.8a	B	The red squirrel is not listed in the Directive and is common throughout Eurasia.
9.9a	A	All populations of the Asian elephant are listed in Appendix I and receive the highest level of protection under CITES. Appendix IV does not exist.
9.10a	D	Protected Birds are listed in the EU Wild Birds Directive.
9.11a	A	Such laws generally protect rare birds of prey that routinely return to the same nest site each year.

9.12a	A	Differential pricing involves charging higher fees for visits to popular areas and lower fees for visits to less popular areas in the hope of redistributing visitor pressure.
9.13a	D	The Parties agreed to keep global temperature rise this century to well below 2°C above preindustrial levels and pursue efforts to limit temperature increase to 1.5°C.
9.14a	B	This is the Agreement Between the Government of Canada and the Government of the United States of America on the Conservation of the Porcupine Caribou Herd. The herd migrates between the territories of the two Parties.
9.15a	A	The Global Environmental Facility was established in 1992 at the end of the Rio Earth Summit to provide funding to tackle the world's most pressing environmental problems.
9.16a	C	This related to air pollution produced by a smelter located in Trail (Canada) that that caused damage in Washington (United States). The case established an important principle of international law.
9.17a	A	The Conference on the Conservation of Nature and Natural Resources in Modern African States was held in Arusha, Tanzania, to discuss the future of African wildlife. It was the first such international conference held in Africa.
9.18a	D	The figure identifies most of the important elements of the Convention.
9.19a	D	Sea otters are not generally considered as part of the marine living resources.
9.20a	C	The Agency is an institution of the European Union. It is based in Copenhagen, Denmark.

Chapter 10 Environmental Assessment, Monitoring and Modelling

10.1f	B	A cannon net is a large net tethered to the ground on one side and carried into the air in an arc by rockets covering birds resting on the ground.
10.2f	C	A biodiversity index measures the number of species and the numbers of individuals in each of these species.
10.3f	B	A clinometer is a handheld device that is pointed at the top a tree from a position on the ground and is used to calculate its height using trigonometry.
10.4f	B	Modern biology courses have moved away from teaching taxonomy. As a result younger biologists are less able to identify animal and plant species than their older colleagues.
10.5f	D	Pitfall traps are essentially smooth-sided pots into which terrestrial invertebrates fall.

10.6f	B	NIMBY in an acronym for 'not in my back yard': not in my neighbourhood.
10.7f	C	This is a large, netted, blind-ended structure containing vegetation. Small birds fly in through the open end and are then captured by hand at the blind end.
10.8f	A	The term 'ambient air' refers to the air in the outside environment.
10.9f	B	EIAs were first introduced in the 1960s in the United States.
10.10f	B	A bioassay measures the effect of a substance by assessing the response of living organisms.
10.11f	A	This is a biodiversity index.
10.12f	D	It is essential that all of these devices record the time that they produce images or record variables if they are to be used in research or monitoring.
10.13f	A	A is correct. The distracters are fictitious.
10.14f	C	400nm – 700nm covers the range of light wavelengths used by photosynthesis. It uses blue light (425-450nm) and red light (600-700nm).
10.15f	B	The decibel scale is logarithmic.
10.16f	B	A datalogger could not sample, count and classify insects.
10.17f	A	This is a laboratory technique for measuring dissolved oxygen in water samples.
10.18f	C	This formula measures the porosity of soil. The distracters are other properties of soils.
10.19f	A	This method measures the mass of microorganisms in the soil.
10.20f	C	An auger is used to remove vertical samples from soil.
10.1i	D	Birds are useful in identifying areas for conservation for all of the reasons given. They are visible (unlike, for example, small mammals), well known (because there are many amateur ornithologists) and ubiquitous.
10.2i	D	Indicator taxa are important in facilitating surveys of rarer taxa.
10.3i	A	BOD and SS are the basic measurements made of the final effluent produced by a wastewater treatment works before it is released to a river.
10.4i	C	An EIA should assess and predict the effects of a project from beginning to completion and also consider the ongoing monitoring required thereafter.
10.5i	C	Pigeon blood has been used as a proxy for the blood of children in studies of lead pollution in Manhattan.
10.6i	A	The Hubbard Brook Study involved comparing rivers in two catchments: one deforested and the other intact.

10.7i	C	This device samples small flying insects passing the top of the tube by sucking them into a container located at the base.
10.8i	A	Tigers tend to use particular paths in the forest and have distinctive markings. The distracters would be more difficult to identify as individuals and live in more expansive environments.
10.9i	C	Primary production in savannahs is closely correlated with rainfall.
10.10i	B	A Leslie matrix may be used to predict the growth of a population using information about the numbers, reproductive rates and survival rates in each age class.
10.11i	C	The result of a Monte Carlo simulation is partly determined by chance so it produces a different prediction each time it is run. The distracters are fictitious.
10.12i	C	Methodological drift is the phenomenon whereby a method evolves with time so that the results it produces cannot easily be compared.
10.13i	D	Strategic Rapid Assessment (SRAPA) is fictitious.
10.14i	C	Fish scales possess rings (like trees) and the number and thickness of these rings indicates the age and growth of the fish respectively.
10.15i	A	LIDAR stands for 'light detection and ranging' or 'light imaging, detection, and ranging'.
10.16i	B	A life cycle assessment considers every impact associated with a product from its 'birth' to its 'death' and disposal.
10.17i	C	Air pollution concentrations at a particular place and point in time depend on the nature of the emission and the atmospheric conditions at the time.
10.18i	B	The pyranometer measures radiation from the sun and nothing else.
10.19i	D	These devices can be used to measure leaf area when the leaf is attached to a plant and when it is not. It takes into account the holes in 'perforated' leaves.
10.20i	C	High sulphur dioxide concentration causes the number of tar spots to decrease but does not affect their average size.
10.1a	D	Tubificid worms possess haemoglobin in their blood and are tolerant to relatively high levels of organic pollution so peak first below the outfall. Chironomid (midge) larvae are less tolerant and *Asellus* peaks last.
10.2a	C	Each index makes a different set of assumptions so the values cannot be compared directly. If comparisons are to be made both sites should be assessed using the same method.
10.3a	B	This is the Shannon-Weiner index.
10.4a	A	This is measured as hectares of a particular habitat on a global basis.

10.5a	B	Beta diversity measures the number of unique species in the sites being compared. Alpha diversity is the number of a species at a particular site.
10.6a	B	Density = number of giraffes/area. Area = $2wL$.
10.7a	C	Gap analysis is used in conservation to identify gaps in networks of protected areas.
10.8a	A	This describes the core area model. The other models are fictitious in this context.
10.9a	B	Stochastic models include an element of chance so they should produce a different outcome from the same initial values each time they are used. Deterministic models produce the same outcome each time.
10.10a	C	Models are only useful in conservation if they are framed with the purpose of informing conservation management from the outset.
10.11a	C	The distracter possesses the same elements in a different (incorrect) order.
10.12a	A	This is a method of assessing plant cover.
10.13a	D	This is a Leslie matrix. The variable f is fecundity, so f_2 is the number of offspring born to an individual in the oldest age group.
10.14a	A	l_x is the number of organisms surviving to age x. The distracters relate to other columns in a life table.
10.15a	C	Calculated as $(2\times5)/(9+8) = 0.59$.
10.16a	C	This curve has a negative exponential form.
10.17a	B	Cluster analysis assembles items into hierarchical groups based on similarities in, and differences between, the same variables measured in each of them. In this example, the method would identify clusters of similar soils based on their characteristics rather than their geographical distribution among the sites.
10.18a	B	A two-way ANOVA examines the effect of two different categorical independent variables (plant species and fertiliser application rate) on one continuous dependent variable (plant growth).
10.19a	D	Participatory risk mapping is correct. In this case information about perceived risk could be gathered from questionnaires or structured interviews and the responses compared with data on the quantity of crops lost to particular species.
10.20a	D	All of these methods may be used to measure LAI.

References

Andermann, T., Faurby, S., Turvey, S.T., Antonelli, A. and Silvestro, D. (2020) The past and future human impact on mammalian diversity. *Science Advances* 6(36), eabb2313. DOI: 10.1126/sciadv.abb2313.

Benkwitt, C.E. (2015) Non-linear effects of invasive lionfish density on native coral reef fish communities. *Biological Invasions* 15, 1383–1395.

Blair, R.B. (1996) Land use and avian species diversity along an urban gradient. *Ecological Applications* 6, 506–519.

Bradshaw, A.D. (1987) The reclamation of derelict land and the ecology of ecosystems. In: Jordan III, W.R., Gilpin, M.E. and Aber, J.D. (eds) *Restoration Ecology*. Cambridge University Press, Cambridge, pp. 53–74.

Diamond, J.M. (1975) The island dilemma: lessons of modern biogeographic studies for the design of nature reserves. *Biological Conservation* 7, 129–146.

Donlan, C.J., Berger, J., Bock, C.E., Bock, J.H., Burney, D.A. *et al.* (2006) Pleistocene rewilding: an optimistic agenda for twenty-first century conservation. *American Naturalist* 168, 660–681.

Hogan, K.A., Larter, R.D., Graham, A.G.C., Arthern, R., Kirkham, J.D. *et al.* (2020) Revealing the former bed of Thwaites Glacier using sea-floor bathymetry: implications for warm-water routing and bed controls on ice flow and buttressing. *The Cryosphere* 14 (9), 2883–2908.

Hounsome, M. (1979) Bird life in the City. In: Laurie, I.C. (ed.) *Nature in Cities*. John Wiley, Chichester, pp. 179–201.

Laurance, W. and Yensen, E. (1991) Predicting the impacts of edge effects in fragmented habitats. *Biological Conservation* 55, 79–92.

Ralls, K. and Ballou, J.D. (2013) Captive Breeding and Reintroduction. In: Levin, S.A. (ed.) *Encyclopedia of Biodiversity*, 2nd edition. Elsevier, San Diego, pp. 662–667.

Savage, A.M., Hackette, B., Guénard, B., Youngsteadt, E.K. and Dunn, R.R. (2014) Fine-scale heterogeneity across Manhattan's urban habitat mosaic is associated with variation in ant composition and richness. *Insect Conservation and Diversity* 8, 216–228.